Reviews of Environmental Contamination and Toxicology

VOLUME 192

Reviews of Environmental Contamination and Toxicology

Editor
David M. Whitacre

VOLUME 192

Springer

Springer
New York: 233 Spring Street, New York, NY 10013, USA
Heidelberg: Postfach 10 52 80, 69042 Heidelberg, Germany

ISBN 978-1-4419-2452-0 e-ISBN 978-0-387-71724-1

ISSN 0179-5953

Printed on acid-free paper.

e-ISBN: 978-0-387-71724-1

springer.com

Foreword

International concern in scientific, industrial, and governmental communities over traces of xenobiotics in foods and in both abiotic and biotic environments has justified the present triumvirate of specialized publications in this field: comprehensive reviews, rapidly published research papers and progress reports, and archival documentations. These three international publications are integrated and scheduled to provide the coherency essential for nonduplicative and current progress in a field as dynamic and complex as environmental contamination and toxicology. This series is reserved exclusively for the diversified literature on "toxic" chemicals in our food, our feeds, our homes, recreational and working surroundings, our domestic animals, our wildlife and ourselves. Tremendous efforts worldwide have been mobilized to evaluate the nature, presence, magnitude, fate, and toxicology of the chemicals loosed upon the earth. Among the sequelae of this broad new emphasis is an undeniable need for an articulated set of authoritative publications, where one can find the latest important world literature produced by these emerging areas of science together with documentation of pertinent ancillary legislation.

Research directors and legislative or administrative advisers do not have the time to scan the escalating number of technical publications that may contain articles important to current responsibility. Rather, these individuals need the background provided by detailed reviews and the assurance that the latest information is made available to them, all with minimal literature searching. Similarly, the scientist assigned or attracted to a new problem is required to glean all literature pertinent to the task, to publish new developments or important new experimental details quickly, to inform others of findings that might alter their own efforts, and eventually to publish all his/her supporting data and conclusions for archival purposes.

In the fields of environmental contamination and toxicology, the sum of these concerns and responsibilities is decisively addressed by the uniform, encompassing, and timely publication format of the Springer triumvirate:

Reviews of Environmental Contamination and Toxicology [Vol. 1 through 97 (1962–1986) as Residue Reviews] for detailed review articles concerned with any aspects of chemical contaminants, including pesticides, in the total environment with toxicological considerations and consequences.

Bulletin of Environmental Contamination and Toxicology (Vol. 1 in 1966) for rapid publication of short reports of significant advances and discoveries in the fields of air, soil, water, and food contamination and pollution as well as methodology and other disciplines concerned with the introduction, presence, and effects of toxicants in the total environment.

Archives of Environmental Contamination and Toxicology (Vol. 1 in 1973) for important complete articles emphasizing and describing original experimental or theoretical research work pertaining to the scientific aspects of chemical contaminants in the environment.

Manuscripts for *Reviews* and the *Archives* are in identical formats and are peer reviewed by scientists in the field for adequacy and value; manuscripts for the *Bulletin* are also reviewed, but are published by photo-offset from camera-ready copy to provide the latest results with minimum delay. The individual editors of these three publications comprise the joint Coordinating Board of Editors with referral within the Board of manuscripts submitted to one publication but deemed by major emphasis or length more suitable for one of the others.

Coordinating Board of Editors

Preface

The role of *Reviews* is to publish detailed scientific review articles on all aspects of environmental contamination and associated toxicological consequences. Such articles facilitate the often-complex task of accessing and interpreting cogent scientific data within the confines of one or more closely related research fields.

In the nearly 50 years since *Reviews of Environmental Contamination and Toxicology* (formerly *Residue Reviews*) was first published, the number, scope and complexity of environmental pollution incidents have grown unabated. During this entire period, the emphasis has been on publishing articles that address the presence and toxicity of environmental contaminants. New research is published each year on a myriad of environmental pollution issues facing peoples worldwide. This fact, and the routine discovery and reporting of new environmental contamination cases, creates an increasingly important function for *Reviews*.

The staggering volume of scientific literature demands remedy by which data can be synthesized and made available to readers in an abridged form. *Reviews* addresses this need and provides detailed reviews worldwide to key scientists and science or policy administrators, whether employed by government, universities or the private sector.

There is a panoply of environmental issues and concerns on which many scientists have focused their research in past years. The scope of this list is quite broad, encompassing environmental events globally that affect marine and terrestrial ecosystems; biotic and abiotic environments; impacts on plants, humans and wildlife; and pollutants, both chemical and radioactive; as well as the ravages of environmental disease in virtually all environmental media (soil, water, air). New or enhanced safety and environmental concerns have emerged in the last decade to be added to incidents covered by the media, studied by scientists, and addressed by governmental and private institutions. Among these are events so striking that they are creating a paradigm shift. Two in particular are at the center of ever-increasing media as well as scientific attention: bioterrorism and global warming. Unfortunately, these very worrisome issues are now super-imposed on the already extensive list of ongoing environmental challenges.

The ultimate role of publishing scientific research is to enhance understanding of the environment in ways that allow the public to be better informed, The term "informed public" as used by Thomas Jefferson in the age of enlightenment conveyed the thought of soundness and good judgment. In the modern sense, being "well informed" has the narrower meaning of having access to sufficient information. Because the public still gets most of its information on science and technology from TV news and reports, the role for scientists as interpreters and brokers of scientific information to the public will grow rather than diminish.

Environmentalism is the newest global political force, resulting in the emergence of multi-national consortia to control pollution and the evolution of the environmental ethic. Will the new politics of the 21st century involve a consortium of technologists and environmentalists, or a progressive confrontation? These matters are of genuine concern to governmental agencies and legislative bodies around the world.

For those who make the decisions about how our planet is managed, there is an ongoing need for continual surveillance and intelligent controls, to avoid endangering the environment, public health, and wildlife. Ensuring safety-in-use of the many chemicals involved in our highly industrialized culture is a dynamic challenge, for the old, established materials are continually being displaced by newly developed molecules more acceptable to federal and state regulatory agencies, public health officials, and environmentalists.

Reviews publishes synoptic articles designed to treat the presence, fate, and, if possible, the safety of xenobiotics in any segment of the environment. These reviews can either be general or specific, but properly lie in the domains of analytical chemistry and its methodology, biochemistry, human and animal medicine, legislation, pharmacology, physiology, toxicology and regulation. Certain affairs in food technology concerned specifically with pesticide and other food-additive problems may also be appropriate.

Because manuscripts are published in the order in which they are received in final form, it may seem that some important aspects have been neglected at times. However, these apparent omissions are recognized, and pertinent manuscripts are likely in preparation or planned. The field is so very large and the interests in it are so varied that the Editor and the Editorial Board earnestly solicit authors and suggestions of underrepresented topics to make this international book series yet more useful and worthwhile.

Justification for the preparation of any review for this book series is that it deals with some aspect of the many real problems arising from the presence of foreign chemicals in our surroundings. Thus, manuscripts may encompass case studies from any country. Food additives, including pesticides, or their metabolites that may persist into human food and animal feeds are within this scope. Additionally, chemical contamination in any manner of air, water, soil, or plant or animal life is within these objectives and their purview.

Manuscripts are often contributed by invitation. However, nominations for new topics or topics in areas that are rapidly advancing are welcome. Preliminary communication with the Editor is recommended before volunteered review manuscripts are submitted.

Summerfield, North Carolina D.M.W.

Table of Contents

Rev Environ Contam Toxicol 192:1–28

Polycyclic Aromatic Hydrocarbons (PAHs) from Coal Combustion: Emissions, Analysis, and Toxicology

Guijian Liu, Zhiyuan Niu, Daniel Van Niekerk, Jian Xue, and Liugen Zheng

Contents

I. Introduction

Coal is a complex heterogeneous mixture of organic and inorganic constituents of allothigenic or authigenic origin. Besides major (>1%) and minor (0.1%–1%) elements in coal, elements such as As, Se, and Hg occur

Communicated by David M. Whitacre.

G.J. Liu, Z.Y. Niu, J. Xue, L.G. Zheng
CAS Key Laboratory of Crust-Mantle Materials and Environment, School of Earth and Space Sciences, University of Science and Technology of China, Hefei 230026, China

G.J. Liu (✉)
Key Laboratory of Loess and Quaternary Geology, Institute of Earth and Environment, CAS, Xi'an, 710075, Shaanxi, China

G.J. Liu, D. Van Niekerk
The Energy Institute, The Pennsylvania State University, University Park, PA 16802, USA

commonly as trace elements (<1,000 ppm) associated with both organic (e.g., polyaromatic hydrocarbons, PAhs) and inorganic matter (Swaine 2000; Liu et al. 1999; Finkelman 1995). PAHs in coal are the major source of organic pollution and may become easily accessible during combustion, coking, pyrolysis, and other coal preparation processes to make the coal consistent in quality and suitable for selling.

Coal is primarily used as a solid fuel to produce heat through combustion. World coal consumption is about 5,800 million short tons annually, of which about 75% is used for electricity production, and while the remainder is used mainly for heating and indoor cooking. China is the largest coal producer and consumer in the world (Liu et al. 1999). It has been estimated that more than 75% of the energy production in China is based on coal (Chen BH et al. 2004), and more than 400 million people in China rely on coal for their domestic energy needs, such as heating and daily cooking. Because of the limited petroleum and natural gas reserves, and the significant coal reserves (1,000 billion t), in China, it is likely that this coal-based, relatively cheap energy structure will continue for the foreseeable future (Ni 2000; Xu et al. 2000; Zhong and Wang 2000; Liu et al. 2004a,b).

Because of incomplete combustion, large amounts of PAHs are formed and emitted during coal usage. The main pathways for PAHs to enter human bodies are by breathing polluted air or through the food chain. Various forms of PAHs in coal or from coal combustion may affect humans, especially indoor coal combustion. Additionally, PAHs in raw coal can reduce or even poison the activity of catalysts in the refinery, making the stockpile of liquid product unstable, such as allochromatic or color change or odor (Chen et al. 2005; Liu and Zheng 2007). Ezatti et al. (2002) estimated that, in 2000, global mortality from indoor air pollution from solid fuels (such as wood, charcoal, crop residues, but mainly coal), was more than 1.6 million persons. This rate of severe global mortality together with combustion-induced diseases such as pneumoconiosis, dermatosis, and other relevant diseases indicate an urgent need to take immediate action against indoor air pollution, especially in developing regions.

Many studies on PAHs extraction and measurement, distribution, species, and emission from coal use, especially for burning, have intensified in recent years, because of a growing awareness of the potential effects of PAHs on the environment, and also as advanced analytical techniques have been developed. In China, there is considerable concern about carbonaceous aerosol emissions from coal combustion because of the potential influence on climate change (Jacobson 2002) and adverse effects on human health (Lighty et al. 2000; Liu and Zheng 2007; Chuang et al. 1992; Chen et al. 2005). Therefore, roughly one-fourth of the global anthropogenic carbonaceous emissions is believed to originate in China, of which about 70% is the result of coal burning (Cooke et al. 1999; Liu et al. 2000).

II. Species of PAHs

The U.S. Environmental Protection Agency (EPA) has identified 1408 hazardous waste sites as the most serious in the nation. These sites make up the National Priorities List (NPL) and are the sites targeted for long-term federal cleanup activities. PAHs have been found in at least 600 of the sites on the NPL. However, the number of NPL sites evaluated for PAHs is not known (USEPA 1997). As EPA evaluates more sites, the number of sites at which PAHs are found may increase. This information is important because exposure to PAHs may cause harmful health effects and because these sites are potential or actual sources of human exposure to PAHs.

Although there is no definitive legislation concerning PAH abatement, the EPA has listed 16 PAHs as priority pollutants, the latest being effective from 1997 (USEPA 1997): these compounds are naphthalene, acenaphthylene, acenaphthene, fluorene, phenanthrene, anthracene, fluoranthene, pyrene, chrysene, benzo[*a*]anthracene, benzo[*b*]fluoranthene, benzo[*k*]fluoranthene, benzo[*a*]pyrene, indeno(1,2,3-*c*,*d*)pyrene, benzo(*g*,*h*,*i*)perylene, and dibenzo(*a*,*h*)anthracene. It is expected that air quality would be improved soon if the emission of these PAHs was controlled.

These 16 PAHs were chosen to be included in this profile because (1) more information is available on these than on the others; (2) they are suspected to be more harmful than some of the others and they exhibit harmful effects that are representative of the PAHs; (3) there is a greater chance that humans will be exposed to these PAHs than to the others; and (4) of all the PAHs analyzed, these were the PAHs identified at the highest concentrations at NPL hazardous waste sites. In China, 9 PAHs, including benzo[*a*]pyrene, were defined as priority PAHs by the State Environmental Protection Administration (Yue et al. 2003).

III. PAH Extraction and Measurement

A. Extraction

Various difficulties exisit in extracting PAH species from coal and coal combustion products. Extraction methods have used traditional methods (Purushothama et al. 1998; Claessens et al. 1991; Chen et al. 1996; Stephens et al. 1994; Leonhardt and stahl 1998; Yu et al. 1995; Hageman et al. 1996; Xue et al. 2007; Lee et al. 1981) such as ultrasonic, soxhlet, and other more recent extraction methods (e.g., supercritical fluid extraction and microwave extraction). Soxhlet extraction is the most common method, using benzene, benzene-methanol, benzene-hexane, methylene chloride, ether, toluene and toluene-methanol. For example, Purushothama et al. (1998) carried out an extraction for analysis of PAHs in coal ash. Their work compared various extraction and quantitation techniques for the determination of adsorbed PAHs in coal ash. Aliquots of fly ash from coal combustion spiked with four PAHs were extracted, using three solvents, three methods, and three gas chromatography/mass spectrometry (GC/MS) methods. The results show the solvents that exert the greatest primary effect are CH_2Cl_2 > toluene > > *o*-xylene. Highest recoveries were obtained using the reflux slurry extraction procedure with CH_2Cl_2 and a relatively fast (20°/min) temperature ramp to 310°. With both CH_2Cl_2 and toluene solvents, ultrasonic-assisted extraction affords the best repeatability.

Chen et al. (1996) used batch, soxhlet, and sonication extraction techniques to evaluate the concentration of PAHs in four soils contaminated with coal tar. They concluded that no consistent differences were observed among the three extraction methods or between the two different chromatographic detection systems. The batch extraction technique was considered the least labor intensive and most cost effective method. The use of GC/flame ionization detection (FID) with certified calibration standards was also found to be adequate for routine analysis of known organic compounds. Stephens et al. (1994) obtained higher recoveries in soxhlet extraction for coal stack ash. However, when soxhlet and ultrasonic extractions were compared for urban atmospheric particulates, they performed the ultrasonic extraction and soxhlet extraction with methanol to analyze recovery of PAHs from coal stack ash and used anthracene, anthracene-9,10-dione, benz(*a*)anthracene, and benz(*a*)anthracene-7,12-dione as adsorbates. Soxhlet extraction yielded consistently higher recoveries of these compounds than ultrasonic extraction from a relatively highcarbon ash (6.17% C). Electron microscopy and bulk carbon analyses revealed significant fragmentation of the carbonaceous particles (notorious for strong affinities for PAH) upon sonication. The freshly exposed surfaces of the carbonaceous particles are thought to adsorb solution-phase PAH during the extraction process, thereby yielding lower recoveries of PAH.

Hageman et al. (1996) used water extraction and conventional extractions. They carried out subcritical (hot/liquid) water extraction by simply placing the sample in an extraction cell, filling the cell with water, and heating the system in an oven. After a 15- to 60-mm extraction, the cell was cooled, the water was removed from the extraction cell, and the solubilized organics analyzed using solid-phase microextraction. Quantitative determinations (recoveries typically ranging from ~60% to 140% compared to conventional solvent extraction) of PAHs from soil and air particulate matter were achieved using a 250°C extraction step and isotopically labeled PAHs as internal standards.

The factors impacting extraction, such as time between 6 and 48 hr (Gohda et al. 1993; Naikwadi et al. 1990) have been reported. To analyze the influence of these factors on extraction of PAHs from coal samples, our lab (Xue et al. 2007; Zhang et al. 2005a,b) carried out experiments for investigating the impacts on the 16 USEPA priority PAHs extracted from raw coal from the Huaibei Mine, China. These were performed with different times, different solvents, and different methods. Major effort was focused on finding proper extraction conditions of PAHs from raw coal that would help in the development of clean coal technologies. In this study, we concluded that extraction time, solvents, and methods will affect the Quantities of PAHs extracted from coal samples. The results showed the following factors: (1) The total amount of EPA priority PAHs varies from low to high with increasing extraction time with ultrasonic-assisted extraction. Thirty minutes should be a suitable extraction time because colloids formed as time passed. Extracts within 24 and 48 hr contain the identical amount of priority PAHs using the soxhlet extraction method. The value at 72 hr is higher. Possibly some basic framework was broken and micropores of the coal enlarged with prolonged immersion. This extraction condition is difficult to achieve as the solvent is mainly water in the environment. It can be concluded that 24 hr is a suitable extraction time for PAHs from raw coal using the soxhlet extraction method. (2) CS_2 is best for extracting low molecular weight PAHs whereas CH_2Cl_2 is better for extracting high molecular weight PAHs. Both are excellent extraction solvents compared to C_6H_{14}. CH_2Cl_2 was found to be a suitable extraction solvent when the toxicity equivalence factor (TEF) was used to characterize the toxic properties of PAHs. (3) Both soxhlet and ultrasonic-assisted extraction results give similar PAH profiles; however, the ultrasound-assisted results are lower, indicating less efficient extraction.

B. Measurement

In earlier studies, analysis of the extracts was performed by column chromatography using a UV detector. Techniques such as packed gas chromatography columns using dexsil and poly(dimethylsiloxane) were used (Hanson et al. 1979; Hauser and Pattison 1972; Colier et al. 1995; Wise et al. 1995). Hauser and Pattison (1972) used the method to analyze the aliphatic fraction of air particulate matter. This method, coupled with aerometric data, enables the first direct determination of *n*-alkanes in air particulate matter and extends analytical capability to the C15–C36 range. They also used this method to examine ambient air samples for particulate matter emanating from coal burning. Colier et al. (1995) studied the exhaust from a direct-injection light-duty diesel engine, which was sampled for PAHs at three constant speeds and at five load settings for each speed, using a total-exhaust solvent-stripping apparatus.

In the 1990s, analytical methods changed significantly. UV/visible absorption spectrometry was utilized as a general technique for analysis of organic particle material in ambient samples (Kister et al. 1996). At the same time, fluorescence spectrometry was developed for PAH sample analysis (Nie et al. 1993). Its main advantage over absorption methods of UV/visible was its high sensitivity by a factor of 10^2–10^3 (Guilbault 1990); its main disadvantage was the interference by trace fluorescent compounds. Qian and Hsu (1992) studied molecular transformation in hydrotreating processes by online liquid chromatography/mass spectrometry. Manoli and Samara (1996) and Wise et al. (1995) performed a modified solid-phase extraction technique using sonication of the adsorbent material instead of the elution normally applied and compared with two conventional liquid-liquid extraction procedures for the determination of the 16 EPA PAHs in municipal waste waters by means of HPLC coupled with fluorescence detection. The best results were obtained by use of acetonitrile.

Cleanup of sewage sludge extracts was not found necessary for accurate quantification of the major PAH components with fluorescence detection. The precision of the whole analytical procedure from extraction to the final determination of PAHs was satisfactory for both wastewater and sewage sludge samples. Murayama and Das Gupta (1996) used liquid chromatography-electrochemical detection (LC-ED) to determine nitro-substituted polynuclear aromatic hydrocarbons that occur in ambient suspended particulate matter. They proposed a sensitive liquid chromatographic method with an unique selectivity: an electrochemical detector operating in the reductive mode is followed by a fluorescence detector. When the NO_2 group is reduced to the NH_2 group, there is a major increase in fluorescence; nitro-PAH compounds are essentially nonfluorescent. A different fluorescence signal or a different chromatofluorogram is generated by subtracting the chromatogram obtained with the electrochemical detector off from that obtained with the electrochemical detector on. Applications to diesel engine exhaust samples are demonstrated.

Carcia et al. (1994) studied cloud point preconcentration and HPLC determination of PAHs with fluorescence detection. Heaton et al. (1994) collected several environmental samples containing PAHs obtained from various sources including soil, water, petrochemicals and coal, and all samples were analyzed by supercritical fluid chromatography (SFC) under optimized conditions. Their conclusion showed that packed-column SFC is shown to be an efficient and promising method for the rapid monitoring of environmental PAHs. Fourier transform infrared spectroscopy was used to determine modifications induced by the preheating process in the distribution of families of PAHs in chloroform and pyridine extracts of two bituminous coals from Spain (Kister et al. 1996). Kovalenko et al. (1992) studied microscopic organic analysis using two-step laser MS, and Chasey and Aczel (1991) used high-resolution MS to determine PAH structure distributions. The other techniques such as supercritical fluid chromatography (Vayisoglu-Giray et al. 1998; Heaton et al. 1994) and capillary electrophoresis with fluorescence detection (Nie et al. 1993) were also used to determine PAHs. Most recent studies use GC/MS to determine PAH species and concentratons (Chen YJ et al. 2004; Xue et al. 2006; Zhang et al. 2005a,b; Wang et al. 1994).

IV. Toxicology and Distribution of PAHs

A. Toxicology

PAHs are highly lipid soluble and are absorbed through the lungs, gut, and skin of mammals. PAHs can easily be associated with fine particles from coal combustion, and the fine ashes can be breathed into the lungs and gut, along with skin exposure. Studies on lung retention of microcrystalline PAHs or PAHs in solution after intratracheal instillation in female rats have indicated that they are rapidly cleared from the respiratory tract (Tornquist 1985). Clearance of the PAHs from the lungs is best described as biphasic. For radiolabeled anthracene, benz[*a*]anthracene, 1-nitropyrene, BaP, 6-nitrobenzo[*a*]pyrene, and dibenzo[*c,g*]carbazole, more than 85% of the initial dose was cleared with a half-time of less than 1 hr. The half-times for clearance of the residual redioacivity (1%–15% of the dose) ranged from 26 to 63 hr (Wolff 1989; Weyand and Bevan 1986). However, inhaled PAHs are predominantly adsorbed on soot particles (Chen et al. 2005). After deposition in the airway, the particles can be eliminated by bronchial clearance. PAHs may be partly removed from the particles during transport on the ciliated mucosa and may penetrate into the bronchial epithelium cells (WHO 1987).

When PAHs are adsorbed on particles, the respiratory uptake rate is lower because the partcles are retained for a long period in the respiratory tract (Tornquist 1985). Sun et al. (1984) concluded that radiolabeled BaP adsorbed on diesel engine exhaust particles was inhaled by rats, lung clearance of the inhaled particle-associated radioactivity occurred in two phases. The initial rapid clearance had a half-time of less than 1 hr. The second, long-term component had a half-time of 18 d and represented 50% of the radioactivity that had initially been deposited in the lungs.

Some studies showed that PAHs are rapidly and widely distributed in the organism because of high lipophilicity (Swartz et al. 1995). For example, some authors (Foth et al. 1988; Modica et al. 1983) reported that the pattern of distribution of BaP was found to be similar after subcutaneous, intraveneous, and intratracheal administration to mice and rats, and highest levels were found in the liver. Mammary and other fatty tissues are significant storage depots for PAHs, but owing to rapid metabolism no significant accumulation seems to take place. Schlede (1970) and Wiersma and Roth (1983) reported that the gastrointestinal tract contains relatively high levels of metabolites as the result of hepatobiliary excretion. Withey et al. (1991, 1992) carried out experiments with mice and showed that the distribution of pyrene in tissues was highest in the perirenal fat, intermediate in the liver, kidneys, and lungs and lowest in the heart, testes, spleen, and brain. Withey et al. (1993) and Neubert and Tapken (1988) concluded that BaP can readily cross the placental barrier of rats and mice, consistent with the fetal and developmental toxicity of the substance.

According to Van de Wiel et al. (1993) and Foth et al. (1988), following metabolism, hepatobiliary excretion and elimination through the feces is the major route by which BaP is removed from the body, independent of the route of administration. Urine is the other major excretory route, although it is quantitatively of minor importance compared to the bile.

Other studies (Hall et al. 1989; Glatt and Oesch 1987; Nebert et al. 1993; Graslund and Jernstrom 1989; Bjelogrlic and Kirsi 1991) showed that the enzyme system primarily responsible for PAH metabolism is the microsomal mixed-function oxidase (MFO) system, which converts the nonpolar PAHs into polar hydroxyl and epoxy derivatives. Epoxides are reactive and enzymatically metabolized to other compounds, such as dihydrodiols and phenols. The enzyme systems that metabolize PAHs are widely distributed in the cells and tissues of humans and animals. The highest metabolizing capacity is present in the liver, followed by lung, intestinal mucosa, skin, brain, hair follicles, erythrocytes, platelets, leukocytes, placenta, and uterus. Animal and human fetal tissues have the capcity to metabolize PAHs, but at a low rate compared to adult tissues.

Some studies (Graslund and Jernstrom 1989; Cavalieri and Rogan 1985; Reddy et al. 1992; Shaw and Connell 1994; Ross et al. 1990, 1991) concluded that PAHs exert their mutagenic and carcinogenic activity through biotransformation to chemically reactive intermediates that bind covalently to cellular macromolecules. Other experiments (Shaw et al. 1994) were carried out to explain the implication of covalent binding of PAHs to DNA for carcinogenicity in the lung and other tissues of experimental animals.

B. Sources and Distribution of PAHs in the Environment

The sources of PAHs are original from combustion processing (Mitra and Bonnie 1995). PAHs are a group of chemicals that are formed during the incomplete burning of coal, oil, gas, wood, garbage, or other organic substances, such as tobacco and charbroiled meat. PAHs generally occur as complex mixtures (e.g., as part of combustion products such as soot), and not as single compounds. PAHs usually occur naturally, but they can be manufactured as individual compounds for research purposes; however, not as the mixtures found in combustion products. As pure chemicals, PAHs generally exist as colorless, white, or pale yellow-green solids. They can have a faint, pleasant odor. A few PAHs are used in medicines and to make dyes, plastics, and pesticides. Others are contained in asphalt used in road construction. They can also be found in substances such as crude oil, coal, coal tar pitch, creosote, and roofing tar. They are found throughout the environment in the air, water, and soil, and can occur in the air, either attached to dust particles or as solids in soil or sediment.

PAHs are formed mainly as a result of pyrolytic processes, especially the incomplete combustion of organic materials during industrial and other human activities, such as processing of coal and crude oil, combustion of natural gas, including for heating, combustion of refuse, vehicle traffic, cooking, and tobacco smoking, as well as in natural processes such as carbonization. However, the main source of emission of PAHs into the air is coal combustion. For example, in the Federal Republic of Germany in 1981, the emissions of BaP were estimated to amount to 18 t: about 30% resulted from coke production, 56% from heating with coal, 13% from motor vehicles, and less than 0.5% from the combustion of heating oil (WHO 1987). In China, emission of PAHs consists of more than 70% from coal heating and other processing. In many developed countries, PAHs enter the environment mostly as releases to air from coal combustion (Baek et al. 1991).

Many PAHs, especially carcinogenic PAHs, are found in all surface soils. Typical concentrations in forest soil range from 5 μg to 100 μg/kg. Substantial amounts of PAHs are transferred to forest soil from vegetation litter because the compounds are adsorded from air on organic matter such as leaves and needles. Rural soil contains carcinogenic PAHs at levels of 10–100 μg/kg originating mainly from atmospheric fallout. For both forest and rural soil, values as high as 1,000 μg/kg may occasionally be found. Metropolitan areas have higher PAH concentrations than forest and agricultural areas because of the many sources of fossil fuel combustion. The majority of urban soil concentrations fall in the 600–3,000 μg/kg range. Higher values near areas of heavy transportation and industrialization are probable. Values of the order of 1,000–3,000 μg/kg are regarded as being in the upper range. However, levels of 8,000–336,000 μg/kg have been reported from road dust (Menzie et al. 1992).

PAHs are found in substantial quantities in some foods, depending on the mode of cooking, preservation, and storage, and are detected in a wide range of meats, fishes, vegetables, and fruits. A Dutch market basket study of dietary components for 18-yr-old males involving determination of 17 different PAHs revealed that all these compounds were detected. The most frequently occurring PAHs were benzo[*b*]fluoranthene, fluoranthene, and benzo[*k*]fluoranthene, in 59%, 48%, and 46% of the samples, respectively. The highest concentration of a single PAH was found for chrysene, at 36 μg/kg in the commodity group sugar and sweets. The mean daily intake of the total PAH fraction analyzed was between 5 and 17 μg/kg. The intake of carcinogenic PAH fraction was roughly half these amounts. The largest contribution to the daily PAH intake came from sugar and sweets, cereals, oils, fats, and nuts (De Vos et al. 1990).

Several studies concluded that most of the above 16 PAHs come from fossil fuel combustion and have high toxicity, and especially a carcinogenic character. According to abundant data in the literature and from experimental results, the most significant health effect to be expected from inhalation exposure to PAHs is an excess risk of lung, gut, and skin cancer, especially lung cancer. In the past, chimney sweeps and tar workers were dermally exposed to substantial amounts of PAHs. There is sufficient evidence that skin cancer in many of these workers was caused by PAHs.

Epidemiological studies in coke oven workers, coal gas workers, and employees in aluminum production plants provide sufficient evidence of the role of inhaled PAHs in the induction of lung cancer. An excessively high rate of lung cancer mortality was found in coke oven workers (Hemminki 1993). The studies showed that the increase in lung cancer cases correlates closely with the time spent working on top of ovens where an average BaP concentration of about 30 mg/m^3 has been detected. It should be noted that all these working environments include chemicals other than PAHs that may also have contributed to the excesses of lung cancer.

The movement and distribution of PAHs in the environment depends on properties such as how easily they dissolve in water and how easily they evaporate into the air. PAHs in general do not easily dissolve in water and have a relatively low solubility in water, but they are highly lipophilic. They are present in air as vapors or attached to surfaces of small solid particles. They can travel long distances before they return to earth in rainfall or by particle settling. Some PAHs evaporate into the atmosphere from surface waters, but most stick to solid particles and settle to the bottom of rivers or lakes. In soils, PAHs are most likely to stick tightly to particles. Some PAHs evaporate from surface soils to air. Certain PAHs in soils also contaminate underground water. The PAH content of plants and animals living on the land or in water can be many times higher than the content of PAHs in soil or water. PAHs can break down to longer-lasting products by reacting with sunlight and other chemicals in the air, generally over a period of days to weeks. Breakdown in soil and water generally takes weeks to months and is caused primarily by microorganisms.

V. Indoor PAHs and their Environmental Effects

During coal burning indoors, a large complex mixture of many components including toxic gases and particulate matter is emitted, and particulate matter is mostly fine particles ("soot") that are coated with a mixture of various toxic chemicals including hazardous trace elements and PAHs (Liu and Zheng 2007; Finlayson-Pitts and Pitts 1986). The fine particles not only cause visibility reduction, acid rain, and climate perturbations, but also cause serious health problems by penetrating and delivering coated chemicals into the human respiratory system. It has been reported that the submicron airborne particles, not the organic chemicals adsorbed onto the particles, are responsible for the tumor response, because the particles overload the lung clearance system (Donaldson et al. 1996, 1998; Chuang et al. 1992).

Smaller particles are considered to be more harmful as they remain in the air for a longer period, while larger ones settle to the ground and are less likely to be inhaled. Smaller particles penetrate into deeper portions of the lungs, and clearing can take a long period of time, months or even years. It has also been reported that most toxic elements are concentrated on the smallest respirable particles (Natusch et al. 1974; Ondov et al. 1982; Liu et al. 2004a,b; Liu and Zheng 2006). Some research (Simcik et al. 1999; Mastral and Callén 2000; Nguren et al. 1999, 2002; Chuang et al. 1992; Yelena et al. 2002; Teschke et al. 1989; McDonald et al. 2000) found PAH emissions had become one of the most important sources of PAHs in the environment during coal utilization processes. Thus, pollutants, including PAHs released from coal utilization processes, are an important retarding factor for China's economic, and social development as well as Chinese health (Chuang et al. 1992).

Domestic combustion is one of the major sources of indoor air pollution in developing countries and has been identified as a serious health hazard affecting hundreds of millions of people, especially women, children, and the elderly. Cooking smoke has been shown to cause respiratory diseases, such as chronic bronchitis, emphysema, expectorative cough, and dyspnea (Huang and Batterman 2000; Monn 2001; Harrison et al. 1996; Linsey et al. 1999; Dickhut et al. 2000; Kavouras et al. 2001; Wolfgang et al. 1997; Simcik et al. 1999; Susan and Pratim 1998; Chuang et al. 1991, 1992; Kingham et al. 2000; Minoia et al. 1997; Schauer et al. 1996; Park et al. 2001; Wilson et al. 1989). For example, Teschke et al. (1989) studied case-control and proportionate mortality in Canada, the U.S., Britain, and Denmark, which show that cooks and other food service workers may have elevated risks of cancers of the nasopharynx, buccal cavity, esophagus, lung, and bladder. A pilot study was done to determine if there might be airborne products of cooking that may be risk factors for these cancers in cooks. Air samples were taken in four restaurants and subsequently analyzed for mutagenicity using the Ames assay and for carcinogens using GC/MS. All four samples taken in the restaurant cooking areas were mutagenic to TA98 without metabolic activation, and two were mutagenic to TA100, also without metabolic activation. The ventilation systems in all four restaurants allow the exposure of cooks to both the air from dining room smoking areas and the volatile products of cooking.

Exposure to unvented indoor cooking smoke may cause cancer, particularly lung cancer. In the rural Xuan Wei County, Yunnan Province, lung cancer mortality rates for women are among the highest in China (Chuang et al. 1991, 1992). Most of these women are nonsmokers, and studies have shown that lung cancer in Xuan Wei is associated with domestic use of smoky coal under nonventilated conditions. Chuang et al. (1991, 1992) collected 10 high-volume filter samples ($<10\,\mu m$) from a home inhabited by a person with lung cancer during cooking periods on four consecutive days and subjected these to soxhlet extraction. This composite sample extract was fractionated on a normal-phase column into 7 fractions. The second fraction was the most active in the bioassay, containing mainly PAHs and alkylated PAHs. The two polar fractions, 6 and 7, were the next most active. The most active PAH fraction was further separated into 11 subfractions, based on the number of aromatic carbons. The results, have shown that the high lung cancer mortality rates are associated with exposure to unvented emissions from the indoor burning of smoky coal for cooking and heating.

The emission of pollutants from domestic cooking is mainly the result of incomplete combustion of fuels and is a function of many variables, including, among others, fuel and stove types, air supply conditions, and the frequency and duration of stove use (Chuang et al. 1992; Liu and Zheng 2006; Liu et al. 2000; Benner et al. 1995; Wolfgang et al. 1991, 1993, 1998; Dubowsky et al. 1999). The common use of low-quality noncommercial fuels (fuel wood, agroresidue, dung cakes, etc.) in rural areas of developing countries may result in high pollutant emissions (Li and Ro 2000; Katagiri et al. 1996; Fischer et al. 2000). Residential stoves may produce higher emissions of some important air pollutants as compared to industrial combustion in many circumstances. The main pollutants in smoke from domestic cookstoves are particulate matter (PM), carbon monoxide (CO), and organic compounds. The latter consist of a wide range of substances. Smith (1987) has identified more than 180 polar, 75 aliphatic, and 225 aromatic compounds in wood smoke, yet there are still many unidentified compounds. Among the organic compounds emitted, of special interest are polycyclic organic matter (POM) and formaldehyde (Godish 1989).

POM is a chemical group that contains two or more benzene rings. One particular set of POM that are known to be carcinogens or mutagens are the PAHs. Most PAHs in the atmosphere are released from the incomplete combustion of organic materials. In the atmosphere, PAHs undergo transformations, and the derivatives are usually more toxic than PAH themselves (Nikolaou et al. 1984), therefore increasing the potential harmful effects on human health. Besides PAHs, the aza- and imino-arenes in POM have also been found to be potentially carcinogenic (Smith 1987). Extracts of smoke samples from domestic combustion sources have demonstrated genotoxic and carcinogenic activities (Smith 1987; McCrillis et al. 1992).

As part of the Integrated Air Cancer Project, the USEPA conducted field emission measurement programs in Raleigh, North Carolina, and Boise, Idaho, to identify potential mutagenic impacts of residential wood burning and motor vehicles on ambient and indoor air. The McCrillis et al. (1992) studies included collection of emission samples from chimneys of wood-burning appliances. Parallel projects were conducted in instrumented woodstove test laboratories to quantify woodstove emissions during operations typical of in-house usage, but under more controlled conditions. Three woodstoves were operated in test laboratories. Over a range of burn rates, burning eastern oak, southern yellow pine, or western white pine. Two conventional stoves were tested at an altitude of 90 m. One of the conventional stoves and a catalytic stove were tested at an altitude of 825 m. Decreasing burn rate increased total particulate emissions from the conventional stoves; the catalytic stove's total particulate emissions were unaffected. There was no correlation of total particulate emissions with altitude, whereas total

PAH emissions were greater at the lower altitude. Mutagenicity of the catalytic stove emissions was greater than conventional stove emissions. Emissions from burning pine were more mutagenic than emissions from oak. The indoor pollution concentration and exposure level depend not only on emission but also on ventilation conditions and on whether cooking stoves are physically isolated from main living areas.

Much of developing Asia's population cook and heat using unvented stoves in poorly designed kitchens. As a result, air pollution levels in Asian homes often exceed World Health Organization standards for ambient outdoor air as well as that for typical indoor levels in developed countries (United Nations Environmental Programme 1991). Human exposure to cooking smoke has received increasing attention in indoor air pollution research (Huang and Batterman 2000; Monn 2001; Harrison et al. 1996; Linsey et al. 1999; Dickhut et al. 2000; Kavouras et al. 2001; Simcik et al. 1999; Susan and Pratim 1998; Chuang et al. 1991; Kingham et al. 2000; Minoia et al. 1997; Schauer et al. 1996; Park et al. 2001; Chuang et al. 1992; Liu and Zheng 2007; Liu et al. 2000; Benner et al. 1995; Wolfgang et al. 1991, 1993, 1997, 1998; Dubowsky et al. 1999). High consumption rates, low quality of fuels, and low efficiency of cookstoves used for domestic cooking in developing countries plus poorly designed kitchens lead to high health risks from cooking smoke. Yet, only limited data on PAH emissions and the toxic effects of cooking smoke are available in these countries (Liu and Zheng 2006; Daisey et al. 1989; Chen et al. 2005; Chuang et al. 1992). Indoor pollution is receiving more and more attention as a result of emissions from coal combustion, especially in China (Liu and Zheng 2007). Thus, methods to clean fuel coal need to be studied and developed in the future.

VI. Emission and Mechanism of PAH Production During Coal Combustion

Coal is derived from plant material such as cellulose, lignin, resins, spores, leaves, stems, and roots under intense heat and pressure. The organic matter of coal is generally regarded as being composed of two complementary structures (Bhattacharyya 1971; Brooks and Glasser 1986; Li and Peter 1996; Li et al. 1995; Liu et al. 2000; Chen et al. 2005). The dominant component is a macromolecular, insoluble, three-dimensional network composed of condensed aromatic and hydroaromatic units connected by ether or thioether linkages and short alkyl bridges. The second structure is a molecular phase of compounds that are typically soluble in organic solvents. This molecular phase contains varying distributions of aliphatic hydrocarbons, polycyclic aromatic and hydroaromatic hydrocarbons, hydroxylated polycyclic aromatic compounds, and heterocyclic compounds.

In the atmosphere, PAHs are mainly associated with aerosols (dispersions or suspensions of solid particulate), liquid drops, or a mixture of both (Smith 1984). The main source of concern with this class of compounds is that some members are known mutagens or carcinogens. PAHs have a strong electrophilic character and interact with biological nucleophyles in metabolic processes. The result of such interactions may obstruct their regular functions and promote carcinogenesis as a result of transformation of PAHs into diolepoxides of aromatic rings inside the organism (Mastral et al. 1996a).

Most PAH formation and emission is associated with soot emissions during the incomplete combustion of fossil fuels, such as the start-up and shut-down process of combustion facilities including troubleshooting, and as a consequence of the pyrolytic process that precedes combustion (Hershkowitz et al. 1983; Mastral et al. 1997a,b). Also, as a consequence of the thermion process and by cyclization and aromatization reactions, other aromatic clusters can be developed (Zander 1980; Bruce et al. 1997). It has been experimentally shown that it is possible to synthesize benzo[*a*]pyrene from ethane under pyrolysis conditions (Li and Peter 1996). When coal is heated, its structure undergoes chemical and physical changes, and organic fractions are released. These fractions undergo cyclization reactions leading to polycyclic compounds that can exist in gaseous and in solid phases, depending on the surrounding temperature and their molecular volume (Gantner et al. 1985; De Wiest and Rondia 1976). They can exist in the gas phase (<-ring PAHs), in the solid phase (>6-ring PAHs), or in both phases (4- and 5-ring PAHs) in the air depending on their molecular mass (Hoffmann et al. 1977). Junk and Ford (1980) found 48 PAHs and Purushothama et al. (1998) found 37 organic compounds that were identified, and 28 of the 37 were PAHs. Some of the these sparingly volatile organics are found in stack gases, but all are found as adsorbates on fly ash and other ash.

PAHs are molecules built up of benzene rings that resemble fragments of single layers of graphite. They have planar structures and a wide variety of shapes and sizes.The dominant component of coal seems to be composed of a three-dimensional network of condensed aromatic and hydroaromatic units connected by weaker bonds (Solomon et al. 1988). During coal use, e.g., heating or combustion, coal structures undergo major physical and chemical changes and release volatile organic compounds. Weaker bonds in aliphatic bridges and rings break up first, then aromatic structures, resulting in more aromatic hydrocarbons in the products. Some of the organic compounds may be emitted as unburned material, thus becoming a source of organic emissions.

According to some (Williams et al. 1986; Barbella et al. 1990; Mastral et al. 1996b, 1999a,b; Purushothama et at. 1998; Chen et al. 2005), during coal combustion PAH formation and emission mechanisms can be classified in two processes: pyrolysis and pyrosynthesis. The have concluded that on coal heating, the organic compounds are partially cracked to smaller and unstable fragments (pyrolysis). These fragments, mainly highly reactive free radicals with a very short average lifetime, lead to more stable PAH formation through recombination reactions (pyrosynthesis). Thus, BaP and other

PAHs are formed through pyrolysis of methane, acetylene, butadiene, and other compounds. Mastral et al. (1997a) reported that PAH formation in combustion is like a waterfall mechanism in which PAH are formed through small radicals to which radicals are added, forming compounds of higher molecular weight, soot, and fullerenes.

Williams et al. (1986) analyzied the polynuclear aromatic compound fractions from five diesel fuels, a gas-oil, and a sample of kerosene. The principal polynuclear aromatic hydrocarbons of the fuels were naphthalene, fluorine, and phenanthrene and their alkyl derivatives; mutagenic compounds were present in significant concentrations. The principal polynuclear aromatic nitrogen and sulfur compounds of diesel fuel are carbazole and dibenzothiophene and their alkyl derivatives. The combustion process formed a significant concentration of mutagenic compounds in the particulate.

In-cylinder sampling and analysis of particulate (soot and condensed hydrocarbon species), light hydrocarbons, and gaseous inorganic species has been performed at two positions of a single-cylinder direct-injection diesel engine by means of a fast sampling valve to follow the behavior of a diesel oil during the engine cycle (Barbella et al. 1990). The combustion process in a diesel engine proceeds through a preignition phase of heating and vaporization of the injected fuel, which causes the transformation of the fuel components in light gaseous products and oxygenated compounds that rapidly disappear at ignition. After ignition, the formation of soot was found, particularly in the high-temperature position, but the formation of soot precursors, i.e., acetylene and combustion-formed PAHs, was not in evidence. In any phase of the combustion cycle, the PAHs are formed from unburned fuel, but in the low-temperature position the PAH percentage increases, as opposed to the high-temperature position, and this can be caused by the high resistance of the fuel PAHs to complete oxidation at this temperature. At the same time, the appearance of high molecular weight material of oxygenated character has been found at the low-temperature condition, this could derive from the oxidation, perhaps in liquid phase, of the heavy aromatic compounds of the fuel. The chemical analysis of the hydrocarbons collected in the oxidation phase and in the exhaust revealed that unburned fuel, reentraining the sampling zones from wall quenching and ring crevices, is the main source of hydrocarbon emission from the engine.

Mastral et al. (1999a,b) studied the PAHs and the organic matter (OM) content associated with particulate matter (PM) emissions from atmospheric fluidized bed coal combustion. The two main aims of the work were (a) to study OM and PAH emissions as a function of the coal fluidized-bed combustion (FBC) variables in solid phase, and (b) to check if there is any correlation between OM and PAH contained in the PM. Mastral et al. (1996b) assessed the influence of FBC temperature and coal rank on PAH emissions. It can be concluded that there a direct relationship between the OM content and the PAH is not supported in the PM emitted. In addition, neither PM or OM show dependence between themselves. A low-rank coal and a high-rank coal were burned in a fluidized-bed reactor (FBR) on a laboratory scale. An outlet gas sample was passed through a capture system provided with three traps: nylon filter, Teflon filter, and XAD-2 resin. The PAHs captured were analyzed by fluorescence spectroscopy after sonic extraction with dimethylformamide. The total PAH emission depends mainly on the pyrolytic process and to a lesser degree on the combustion efficiency. Although the total amount of PAH emitted follows a specific trend as a function of the combustion temperature, the amount of each PAH emitted seems to be the consequence of random distribution from PAH interconversion and association. Under all conditions of combustion, the amount of PAH emitted in the gas phase is higher than that collected in two cyclones preceding the gas sampling location.

Other papers reported the mechanism of emission and formation during coal or fuel combustion. Mastral et al. (1995) took into account the mechanism of PAH formation and showed that, although the individual PAH amount found in atmospheric fluidized-bed combustion (AFBC) as a function of the combustion temperature follows a random distribution, the total of PAH maintains a trend. Visser et al. (1998) studied the PAH rearrangement and interconversion processes during combustion, showing that the formation of CP-PAH from E-PAH as well as CP-PAH rearrangements and/or interconversions are readily recognized. GC-infrared (IR), being complementary to GC-MS, enables the identification and structural assignment of previously undetected isomeric PAHs, which are only present in trace-level amounts. Williams and Taylor (1993) concluded that the results are of interest for the elucidation of PAH buildup processes and for the rationalization of the ubiquitous formation of a distinct set of PAHs during combustion.

Acevedo et al. (1996) concluded that the complex hydrocarbons do not have to necessarily break into small fragments before performing recombination processes. Compounds with several rings can undergo partial cracking followed by dehydrogenation of the primary radicals. References have been found in our review suggesting that lipid acids with a long linear chain play an important role in the PAH formation mechanism because they allow the alkylation of aromatic units in fossil fuels. These calculations give theoretical support to experimental evidence in the literature suggesting that massive systems, more than six condensed aromatic rings, are not likely to be found in significant quantities in petroleum samples.

On the other hand, Maximilian and Gerd (1993) suggested that intermolecular and intramolecular hydrogen transfer as well as phenyl radicals play an important role as intermediary compounds in high-temperature reactions that lead to polycyclic compound formation. As PAHs and structurally related hetero-aromatic systems are the main constituents of pitches, there are many relationships between polycyclic aromatic chemistry and pitch science. Lewis (1982) carried out an experiment on thermal polymerization of aromatic hydrocarbons, and concluded that, in combustion processes, tars are also produced as a consequence of condensation of released emissions, and the typical reactions of PAH formation in tars have been explained through dehydrogenating polymerization processes together with an aromatization process, or aromatization growth. Fragmentation reactions, resulting from hydrogen transfer and posterior division of simple C–C bonds formed initially, and alkylation reactions have also been observed by Mastral et al. (1997b). Fetzer and Kershaw (1995), using HPLC with a shape-selective reversed phase and a diode array absorbance detector, identified several 7- to 10-ring PAHs in coal tar pitches. Their structures suggest that many different reaction pathways occur in the production of coal tar pitches.

VII. Factors Influencing Formation and Emission

A. Coal Quality

According to our results (Zhang et al. 2005a,b; Xue et al. 2006, 2007), the coal rank and the levels of carbon, hydrogen, oxygen, sulfur, and coal structure will affect the species and contents of PAHs in coal. That is, coal characteristics are important factors that control the amount and species of PAHs in coal. The samples of Huaibei coal seam No. 4 and coal seam No. 5, led us to conclude the following. (1) The organic compounds in No. 4 and No. 5 coals from Huaibei Coalfield are mainly dibenzothiophene, benzo[*b*]naphtho[2,3-*d*]thiophene, and their methyl and dimethyl derivatives and their isomers. A larger number of PAH species occur in the No. 5 coal than in the No. 4 coal. The relative abundance of individual PAHs in samples of the No. 5 coal correlates negatively with the H/C and O/C ratios, indicating a possible relationship to metamorphic grade. (2) Species and abundance of organic sulfur compounds were greater in coal seam No. 5 than those in coal seam No. 4, because coal seam No. 5 has a higher degree of coalification as a result of magmatic intrusion. (3) When total sulfur is about 0.5%, the total relative content of organic compounds is largest, because the organic sulfur is mainly from coal-forming plants without the effect of marine sedentary environments.

B. Combustion Conditions

PAHs are the major products formed and emitted from coal burning. Combustion conditions are the most influential factor on formation and emission of PAHs from coal. In general, the major factors influencing combustion conditions are combustor type (such as indoor coal stove combustion, fluidized-bed combustion) and combustion parameters (such as temperature, level of oxygen), among others (Mastral and Callén 2000; Hayashi et al. 1993; Purushothama et al. 1998,). For example, the most common and ancient combustion practice to produce energy still used is domestic coal burning for heating (in residential stoves). Knobloch and Engewald (1993) concluded that PAH emissions from this use were higher than in the other two controlled systems (industrial stokers and fluidized-bed combustion burning). Spuznar (1992) compared PAH emissions from three coal-fired power stations (pulverized coal/oil; burning of low-sulfur coals with electrostatic precipitators; with electrostatic precipitators plus a fuel gas desulfurization unit) and detected the lowest emissions for the largest unit. Regarding temperature, the highest PAH emissions were produced in the range between 750° and 850°C. For the percentage of excess air (Mastral et al. 1998; Mastral and Callén 2000), it was concluded that a great decrease in PAH emissions was observed when the percentage of excess air was increased from 5% to 10%. The airflow total also made a major contribution to the total of PAHs emitted (Mastral et al. 1999c). Thus, the lowest PAH emissions were emitted at flows double that of the minimum fluidization velocity. At high flows, a deformation in the bed remarkably increased PAH emissions. Other combustion parameters, including combustion temperature, particle size, moisture content, and secondary air, have a major influence on PAH emissions.

Summary

Coal may become more important as an energy source in the 21st century, and coal contains large quantities of organic and inorganic matter. When coal burns chemical and physical changes take place, and many toxic compounds are formed and emitted. Polycyclic aromatic hydrocarbons (PAHs) are among those compounds formed and are considered to pose potential health hazards because some PAHs are known carcinogens. Based on their toxicology, 16 PAHs are considered as priority pollutants by the USEPA.

More attention must be given to the various methods of extraction and analysis of PAH from coal or coal products to accurately explain and determine the species of PAHs. The influences of the extraction time, solvents, and methods for PAH identification are important. In the future, more methods and influences will be studied more carefully and widely.

PAHs are environmental pollutants, are highly lipid soluble, and can be absorbed by the lungs, gut, and skin of mammals because they are associated with fine particles from coal combustion. More attention is being given to PAHs because of their carcinogenic and mutagenic action. We suggest that when using a coal stove indoors, a chimney should be used; the particles and gas containing PAHs should be released outdoors to reduce the health hazard, especially in Southwest China.

During coal utilization processes, such as coal combustion and pyrolysis, PAHs released may be divided into two categories according to their formation pathways: one pathway is derived from complex chemical reactions and the other is from free PAHs transferred from the original coal. The formation and emission of PAHs is a complex physical and chemical process that has received considerable attention in recent years. It is suggested that the formation mechanisms of PAHs will be an increasingly important topic for researchers to find methods for controlling emissions during coal combustion.

Acknowledgments

This work was supported by the National Natural Science Foundation of China (40273035). We thank Prof. C. Zhang for providing medical data, and Professor H. H. Schobert and the Editors for giving much constructive comments and improving the quality of the paper.

References

Acevedo S, Antonieta MR, Luis BG, Escobar G (1996) The PMO method for analysis of structural features of polycyclic aromatic hydrocarbons relevant to asphaltenes. Fuel 75(9):1139–1144.

Baek SO, Field RA, Goldstone ME, Kirk PW, Lester JN, Perry R (1991) A review of atmospheric polycyclic aromatic hydrocarbons: sources, fate and behavior. Water Air Soil Pollut 60(3–4):279–300.

Barbella R, Bertoli C, Ciajolo A, Danna A (1990) Behavior of a fuel oil during the combustion cycle of a direct injection diesel engine. Combust Flame 82(2): 191–198.

Benner BAJ, Wise SA, Currie LA, Klouda GA, Klinedinst DB, Zweidinger RB, Stevens RK, Lewis CW (1995) Distinguishing the contributions of residential wood combustion and mobile source emissions using relative concentrations of dimethylphenanthrene isomers. Environ Sci Technol 29(9):2382–2389.

Bhattacharyya KK (1971) Role of sorption of water vapor in the spontaneous heating of coal. Fuel 50(4):367–380.

Bjelogrlic N, Kirsi V (1991) Benzo[*a*]pyrene-globin adducts detected by synchronous fluorescence spectrophotometry: method development and relating to lung DNA adducts in mice. Carcinogenesis 12(12):2205–2209.

Brooks K, Glasser D (1986) A simplified model of spontaneous combustion in coal stockpiles. Fuel 65(8):1035–1041.

Bruce CR, Wilkinson BB, Culross CC, Holmes SM, Martinez LE (1997) Hydrogen transfer induced cleavage of biaryl bonds. Energy Fuels 11(1):61–75.

Carcia C, Perez JL, Moreno B (1994) Cloud point preconcentration and high performance liquid chromatographic determination of polycyclic aromatic hydrocarbons with fluorescence detection. Anal Chem 66(6):874–881.

Cavalieri E, Rogan E (1985) Role of radical cations in aromatic hydrocarbon carcinogenesis. Environ Health Perspect 64:69–84.

Chasey KL, Aczel T (1991) Polycyclic aromatic structure distributions by high-resolution mass spectrometry. Energy Fuels 5:386–394.

Chen BH, Hong CJ, Kan HD (2004) Exposures and health outcomes from outdoor air pollutants in China. Toxicology 198:291–300.

Chen CS, Suresh PC, Linda SL (1996) Evaluation of extraction and detection methods for determining polynuclear aromatic hydrocarbons from coal tar contaminated soils. Chemosphere 32(6):1123–1132.

Chen YJ, Bi XH, Mai BX, Sheng GY, Fu JM (2004) Emission characterization of particulate/gaseous phases and size association for polycyclic aromatic hydrocarbons from residential coal combustion. Fuel 83(7–8):781–790.

Chen YJ, Sheng GY, Bi XH, Feng YL, Mai BX, Fu JM (2005) Emission factors for carbonaceous particles and polycyclic aromatic hydrocarbons from residential coal combustion in China. Environ Sci Technol 39(6):1861–1867.

Chuang JC, Mack GA, Kuhlman MR, Wilson NK (1991) Polycyclic aromatic hydrocarbons and their derivatives in indoor and outdoor air in an eight-home study. Atmos Environ Part B Urban Atmos 25B(3):369–380.

Chuang JC, Wise SA, Cao SR, Judy LM (1992) Chemical characterization of mutagenic fractions of particles from indoor coal combustion: a study of lung cancer in Xuan Wei, China. Environ Sci Technol 26(5):999–1004.

Claessens HA, Rhemrev MM, Wevers JP, Janssen AAJ, Brasse LJ (1991) Comparison of extraction methods for the determination of polycyclic aromatic hydrocarbons in soot samples. Chromatographia 31(11–12):569–574.

Collier AR, Rhead MM, Trier CJ, Bell MA (1995) Polycyclic aromatic compound profiles from a light-duty direct-injection diesel engine. Fuel 74(3):362–367.

Cooke WF, Liousse C, Cachier H, Feichter J (1999) Construction of a 1° × 1° fossil fuel emission data set for carbonaceous aerosol and implementation and radiative impact in the ECHAM4 model. J Geophys Res 104(D18):22137–22162.

Daisey JM, Spengler JD, Kaarakka P (1989) A comparison of the organic chemical composition of indoor aerosols during wood burning and non-wood burning periods. Environ Int 15(1–6):435–442.

De Vos RH, Van Dokkum W, Schouten AP, De Jong-Berkhout P (1990) Polycyclic aromatic hydrocarbons in Dutch total diet samples (1984–1986). Food Chem Toxicol 28(4):263–268.

De Wiest F, Rondia D (1976) On the validity of determinations of benzo[*a*]pyrene in airborne particles in the summer months. Atmos Environ 10(6):487–489.

Dickhut RM, Canuel EA, Gustafson KE, Liu K, Arzayus KM, Walker SE, Edgecombe G (2000) Automotive sources of carcinogenic polycyclic aromatic hydrocarbons associated with particulate matter in the Chesapeake Bay Region. Environ Sci Technol 34(21):4635–4640.

Donaldson K, Beswick PH, Gilmour PS (1996) Free radical activity associated with the surface of particles: a unifying factor in determining biological activity? Toxicol Lett 88(1–3):293–298.

Donaldson K, Li XY, MacNee W (1998) Ultrafine (nanometer) particle mediated lung injury. J Aerosol Sci 29(5/6):553–560.

Dubowsky SD, Wallace LA, Buckley TJ (1999) The contribution of traffic to indoor concentrations of polycyclic aromatic hydrocarbons. J Expos Anal Environ Epidemiol 9(4):312–321.

Ezzati M, Lopez AD, Rodgers A, Vander HS, Murray CJ (2002) Comparative risk assessment collaborative group of elected major risk factors and global and regional burden of disease. Lancet 360:1347–1360.

Fetzer C, Kershaw JR (1995) Identification of large polycyclic aromatic hydrocarbons in a coal tar pitch.Fuel 74(10):1533–1536.

Finkelman RB (1995) Modes of occurence of environmentally sensitive trance elements in coal. In: Swaine DJ, Goodarzi F (eds) Environmental Aspects of Trace Elements in Coal, Kluwer, Dordrecht, pp 24–50.

Finlayson-Pitts BJ, Pitts JN (1986) Atmospheric Chemistry: Fundamentals and Experimental Techniques. Wiley, & New York, p 12.

Fischer PH, Hoek G, Van Reeuwijk H, Briggs DJ, Lebret E, Van Wijnen JH, Kingham S, Elliott PE (2000) Traffic-related differences in outdoor and indoor concentrations of particles and volatile organic compounds in Amsterdam. Atmos Environ 34(22):3713–3722.

Foth H, Kahl R, Kahl GF (1988) Pharmacokinetics of low doses of benzo[*a*]pyrene in the rat. Food Chem Toxicol 26:45–51.

Gantner E, Steinert D, Reinhardt J (1985) Raman measurements of tributyl phosphate after adsorption on silver hydrosols. Anal Chem 57(8):1658–1662.

Glatt HR, Oesch F (1987) Species differences in enzymes controlling reactive epoxides. Arch Toxicol (Suppl) 10:111–124.

Godish T (1989) Indoor Air Pollution Control. TD883.1.G63. Lewis, Pearl River, NY.

Gohda H, Hatano H, Hanai T, Miyaji K, Takahashi N, Sun Z, Dong Z, Yu H, Cao T, Albrecht ID (1993) GC and GC-MS analysis of polychlorinated dioxins, dibenzofurans and aromatic hydrocarbons in fly ash from coal-burning works. Chemosphere 27(1–2):9–15.

Graslund A, Jernstrom B (1989) DNA–carcinogen interaction: covalent DNA-adducts of benzo[*a*]pyrene 7,8-dihydrodiol 9,10-epoxides studied by biochemical and biophysical techniques. Quant Rev Biophys 22:1–37.

Guilbault GG (1990) Practical Fluorescence, 2nd Ed. Dekker, New York, pp 22–26.

Hageman KJ, Laurent M, Grabanski CB, David JM, Steven BH (1996) Coupled subcritical water extraction with solid-phase microextraction for determining semivolatile organics in environmental solids. Anal Chem 68:3892–3898.

Hall M, Lesley MF, Deborah KP, Grover PL, Roland CW (1989) Relative contribution of various forms of cytochrome P450 to the metabolism of benzo[*a*]pyrene by human liver microsomes. Carcinogenesis 10(10):1815–1821.

Hanson RL, Carpenter RL, Newton GJ, Rothenberg SJ (1979). The human to exposure to PAHs. J Environ Sci Health A14(4):223.

Harrison RM, Smith DJT, Luhana L (1996) Source apportionment of atmospheric polycyclic aromatic hydrocarbons collected from an urban location in Birmingham, U.K. Environ Sci Technol 30(3):825–832.

Hauser TR, Pattison JN (1972) Analysis of aliphatic fraction of air particulate matter. Environ Sci Technol 6:549–557.

Hayashi J, Kawakami T, Taniguchi T, Kusakabe K, Morooka S, Yumura M (1993) Control of molecular composition of tar by secondary reaction in fluidized-bed pyrolysis of a subbituminous coal. Energy Fuels 7:57–66.

Heaton DM, Bartle KD, Clifford AA, Myers P, King BW (1994) Rapid separation of polycyclic aromatic hydrocarbons by packed column supercritical fluid chromatography. Chromatographia 39(9/10):607–611.

Hemminki K (1993) DNA adducts, mutations and cancer. Carcinogenesis 14:2007–2012.

Hershkowitz F, Olmstead WN, Rhodes RP, Rose KD (1983) Molecular mechanism of oil shale pyrolysis in nitrogen and hydrogen atmospheres. In: Geochemistry and chemistry of Oil Shales. ACS Symposium Series 230. Acs, Washington, DC, pp 301–316.

Hoffmann D, Wynder EL (1977) In: Air Pollution, 3rd Ed., vol. II. Academic Press, New York.
Huang YL, Batterman S (2000) Residence location as a measure of environmental exposure: a review of air pollution epidemiology studies. J Expos Anal Environ Epidemiol 10(1):66–85.
Jacobson MZ (2002) Control of fossil-fuel particulate black carbon and organic matter, possibly the most effective method of slowing global warming. J Geophys Res 107:4410.
Junk GA, Ford CS (1980) A review of organic emissions from selected combustion processes. Chemosphere 9(4):187–230.
Katagiri AM, Tamura KK, Yamamoto S, Matsumoto M, Li YF, Cao SR, Ji RD, Liang CK (1996) Indoor and outdoor air pollution in Tokyo and Beijing supercities. Atmos Environ 30(5):695–702.
Kavouras IG, Petros K, Manolis T, Evaggelia L, Euripides GS, Dietrich VB (2001) Source Apportionment of urban particulate aliphatic and polynuclear aromatic hydrocarbons (PAHs) using multivariate methods. Environ Sci Technol 35(11): 2288–2294.
Kingham S, David B, Elliott P, Fischer P, Lebret E (2000) Spatial variations in the concentrations of traffic-related pollutants in indoor and outdoor air in Huddersfield, England. Atmos Environ 34(6):905–916.
Kister J, Pieri N, Alvarez R, Diez MA, Pis JJ (1996) Effects of preheating and oxidation on two bituminous coals assessed by synchronous UV fluorescence and FTIR spectroscopy. Energy Fuels 10:948–957.
Knobloch T, Engewald W (1993) Identification of some polar polycyclic compounds in emissions from brown-coal-fired residential stoves. J High Resolut Chromatogr 16(4):239–242.
Kovalenko LJ, Maechling CR, Clement SJ, Philippoz JM, Zare RN, Alexander CM (1992) Microscopic organic analysis using two-step laser mass spectrometry: application to meteoritic acid residues. Anal Chem 64:682–690.
Lee ML, Novotny M, Bartle KD (1981) Analytical Chemistry of Polycyclic Aromatic Compounds. Academic Press, New York.
Leonhardt E, Stahl R (1998) Decomposition of acenaphthylene by ultrasonic irradiation. Anal Chem 70:1228–1230.
Lewis IC (1982)Thermal polymerization of aromatic hydrocarbons. 18 Carbon (3):191–196.
Li CS, Ro YS (2000) Indoor characteristics of polycyclic aromatic hydrocarbons in the urban atmosphere of Taipei. Atmos Environ 34(4):611–620.
Li CZ, Peter FN (1996) Fate of aromatic ring systems during thermal cracking of tars in a fluidized-bed reactor. Energy Fuels 10(5):1083–1090.
Li CZ, Wu F, Xu B, Kandiyoti R (1995) Characterization of successive time/temperature-resolved liquefaction extract fractions released from coal in a flowing-solvent reactor. Fuel 74(1):37–45.
Lighty JS, Veranth JM, Sarofim AF (2000) Combustion aerosols: factors governing their size and composition and implications to human health. J Air Waste Manag Associ 50(9):1565–1622.
Linsey MC, Kirchstetter TW, Robert AH (1999) Characterization of polycyclic aromatic hydrocarbons in motor vehicle fuels and exhaust emissions. Environ Sci Technol 33(18):3091–3099.

Liu GJ, Zheng LG (2007) Health effects of arsenic, fluorine and selenium from indoor burning of Chinese coal. Rev Environ Contam Toxicol 189:89–106.

Liu GJ, Wang GL, Zhang W (1999) Study on Environmental Geochemistry of the Trace and Minor elements in Coal. China University of Mining and Technology Press, Jiangsu.

Liu GJ, Yang PY, Peng ZC, Chou CL (2004a) Petrographical and geochemical contrasts and environmentally significant trace elements in marine-influenced coal seams, Yanzhou Mining Area, China. J Asian Earth Sci 23(4):491–506.

Liu GJ, Zhang HY, Gao LF, Zheng LG, Peng ZC (2004b) Petrological and mineralogical characterizations and chemical composition of coal ashes from power plants in Yanzhou Mining District, China. Fuel Process Technol 85(15):1635–1646.

Liu KL, Xie W, Zhao ZB, Pan WP, John TR (2000) Investigation of polycyclic aromatic hydrocarbons in fly ash from fluidized bed combustion systems. Environ Sci Technol 34(11):2273–2279.

Manoli E, Samara C (1996) Polycyclic aromatic hydrocarbons in waste waters and sewage sludge: extraction and clean-up for HPLC analysis with fluorescence detection. Chromatographia 43(3/4):135–142.

Mastral AM, Callén MS (2000) A review on polycyclic aromatic hydrocarbon (PAH) emissions from energy generation. Environ Sci Technol 34(15):3051–3056.

Mastral AM, Callén MS, Mayoral C, Galbán J (1995) Polycyclic aromatic hydrocarbon emissions from fluidized bed combustion of coal. Fuel 74(12):1762–1766.

Mastral AM, Murillo R, PerezSurio MJ, Perez-Surio, Callén MS (1996a) Coal hydrocoprocessing with tires and tire components. Energy Fuels 10(4):941–947.

Mastral AM, Callén MS, Murillo R (1996b) Assessment of PAH emissions as a function of coal combustion variables. Fuel 75(13):1533–1536.

Mastral AM, Callen MS, Murillo R, Mayoral C (1997a) Proceedings, 9th International Conference on Coal Science, Essen, Germany, September, pp 7–12.

Mastral AM, Murillo R, José MP, Mayoral C, Callén M (1997b) Iron-catalyzed coal-tire coprocessing: influence on conversion products distribution. Energy Fuels 11(4):813–818.

Mastral AM, Callén MS, Murillo R, García T (1998) Assessment of PAH emissions as a function of coal combustion variables in fluidised bed. 2. Air excess percentage. Fuel 77(13):1513–1516.

Mastral AM, Callén MS, Murillo R, García T (1999a) Influence on PAH emissions of the air flow in AFB coal combustion. Fuel 78(13):1553–1557.

Mastral AM, Callén MS, Murillo R, García T (1999b) Combustion of high calorific value waste material: organic atmospheric pollution. Environ Sci Technol 33(23): 4155–4158.

Mastral AM, Callén MS, García T (1999c) Polycyclic aromatic hydrocarbons and organic matter associated to particulate matter emitted from atmospheric fluidized bed coal combustion. Environ Sci Technol 33(18):3177–3184.

Maximilian Z, Gerd C (1993) A review of the significance of polycyclic aromatic chemistry for pitch science. Fuel 72(9):1281–1285.

McCrillis RC, Watts RR, Warren SH (1992) Effects of operating variables on PAH emissions and mutagenicity of emissions from woodstoves. J Air Waste Manag Assoc 42(5):691–694.

McDonald JD, Barbara Z, Fujita EM, Sagebiel JC, Chow JC, Watson JG (2000) Fine particle and gaseous emission rates from residential wood combustion. Environ Sci Technol 34(11):2080–2091.

Menzie CA, Potocki BB, Santodonato J (1992) Exposure to carcinogenic PAHs in the environment. Environ Sci Technol 26:1278–1284.

Minoia C, Magnaghi S, Micoli G, Fiorentino ML, Turci R, Angeleri S, Berri A (1997) Determination of environmental reference concentration of six PAHs in urban areas (Pavia, Italy). Sci Total Environ 198(1):33–41.

Mitra S, Bonnie R (1995) Patterns and sources of polycyclic aromatic hydrocarbons and their derivatives in indoor air. Atmos Environ 29(22):3345–3356.

Modica R, Fiume M, Guaitani A, Bartosek I (1983) Comparative kinetics of benz[*a*]anthracene, chrysene and triphenylene in rats after oral administration. I. Study with single compounds. Toxicol Lett 18(1–2):103-109.

Monn C (2001) Exposure assessment of air pollutants: a review on spatial heterogeneity and indoor/outdoor/personal exposure to suspended particulate matter, nitrogen dioxide and ozone. Atmos Environ 35(1):1–32.

Murayama M, Dasgupta PK (1996) Liquid chromatographic determination of nitro-substituted polynuclear aromatic hydrocarbons by sequential electrochemical and fluorescence detection. Anal Chem 68(7):1226–1232.

Naikwadi KP, Karasek FW, Hatano H (1990) Analyses of polychlorinated dibenzo-*p*-dioxins and dibenzofurans and precursors in fly ash samples collected at different points in post-combustion zone of Japanese machida incinerator. J Chromatogr A 511(1):281–290.

Natusch DFS, Wallace JR, Evans CAJ (1974) Toxic trace elements. Preferential concentration in respirable particles, Science 183(4121):202–204.

Nebert DW (1993) Role of the Ah receptor and the dioxin-inducible{Ah} gene battery in toxicity, cancer, and signal transduction. Ann N Y Acade Scie 685:624–640.

Neubert D, Tapken S (1988) Transfer of benzo[*a*]pyrene into mouse embryos and fetuses. Arch Toxicol 62(2–3):236–239.

Nguyen TKO, Lars BR, Nghiem TD (1999) Emission of polycyclic aromatic hydrocarbons and particulate matter from domestic combustion of selected fuels. Environ Sci Technol 33(16):2703–2709.

Nguyen TKO, Le HN, Yin LP (2002) Emission of polycyclic aromatic hydrocarbons, toxicity, and mutagenicity from domestic cooking using sawdust briquettes, wood, and kerosene. Environ Sci Technol 36(5):833–839.

Ni B (2000) New progress in high-precision and high resolution seismic exploration technology in coal industry of China. Acta Geol Sin 74(2):311–314.

Nie S, Dadoo R, Zare RN (1993) Ultrasensitive fluorescence detection of polycyclic aromatic hydrocarbons in capillary electrophoresis. Anal Chem 65(24):3571–3575.

Nikolaou K, Masclet P, Mouvier G (1984) Sources and chemical reactivity of polynuclear aromatic hydrocarbons in the atmosphere: a critical review. Sci Total Environ 32(2):103–132.

Ondov JM, Zoller WH, Gordon GE (1982) Trace element emissions in aerosols from motor vehicles. Environ Sci Technol 16(6):318–328.

Park JS, Wade TL, Sweet S (2001) Atmospheric distribution, of polycyclic aromatic hydrocarbons and deposition to Galveston Bay, Texas, USA. Atmos Environ 35(19):3241–3249.

Purushothama S, Lioyd EG (1994) PAHs in coals. J Coal Qual 13:77–79.

Purushothama S, Pan WP, Riley JT, Lloyd WG (1998) Analysis of polynuclear aromatic hydrocarbons from coal fly ash. Fuel Process Technol 53(3):235–242.

Qian K, Hsu CS (1992) Molecular transformation in hydrotreating processes studied by on-line liquid chromatography/mass spectrometry. Anal Chem 64:2327–2333.

Reddy AP, Donna PS, Chuan J, Peter G, Lawrence JM (1992) Phosphorus-32-postlabeling analysis of DNA adduction in mouse skin following topical administration of (+)-7, 8-dihydroxy-7,8-dihydrobenzo[*a*]pyrene. Chem Res Toxicol 5(1):26–33.

Ross J, Nelson G, Kligerman A, Erexson G, Bryant M, Earley K, Gupta R, Nesnow S (1990) Formation and persistence of novel benzo(a)pyrene adducts in rat lung, liver, and peripheral blood lymphocyte DNA. Cancer Res 50(8):5088–5094.

Ross J, Nelson G, Erexson G, Kligerman A, Earley K, Gupta R, Nesnow S (1991) DNA adducts in rat lung, liver and peripheral blood lymphocytes produced by i.p. administration of benzo[*a*]pyrene metabolites and derivatives. Carcinogenesis 12(10):1953–1955.

Schauer JJ, Wolfgang RF, Hildemann LM, Mazurek MA, Glen CR (1996) Source apportionment of airborne particulate matter using organic compounds as tracers. Atmos Environ 30(22):3837–3855.

Schlede E (1970) Stimulatory effect of benzo[alpha]pyrene and phenobarbital pretreatment on the biliary excretion of benzo[alpha]pyrene metabolites in the rat. Cancer Res 30(6):1898–1908.

Shaw GR, Connell DW (1994) Prediction and monitoring the carcinogenicity of polycyclic aromatic compounds (PACs) Rev Environ Contam Toxicol 135: 1–62.

Simcik MF, Eisenreich SJ, Lioy PJ (1999) Source apportionment and source/sink relationships of PAHs in the coastal atmosphere of Chicago and Lake Michigan. Atmos Environ 33(30):5071–5079.

Smith I (1984) PAH from Coal Utilization Emissions and Effects. Coal Research Report. IEA, London.

Smith KR (1987) Biofuels, Air Pollution, and Health: A Global Review. Plenum Press, New York.

Solomon PR, Hamblen DG, Carangelo RM, Serio MA, Deshpande GV (1988) General model of coal devolatilization. Energy Fuels 2(4):405–422.

Spuznar CB (1992) Air toxic emissions from the combustion of coal: identifying and quantifying hazardous air pollutants from US coals. ANL/EAIS/TM-83. Argonne National Laboratory, Environmental Assessment and Information Sciences Division, Argonne, IL.

Stephens DL, McFadden T Jr, Heath DO, Mauldin RF (1994) The effect of sonication on the recovery of polycyclic aromatic hydrocarbons from coal stack ash surfaces. Chemosphere 28(10):1741–1747.

Sun JD, Wolff RK, George MK, McClellan RO (1984) Lung retention and metabolic fate of inhaled benzo(a)pyrene associated with diesel exhaust particles. Toxicol Appl Pharmacol 73(1):48–59.

Susan KD, Pratim B (1998) Characterization of polycyclic aromatic hydrocarbon particulate and gaseous emissions from polystyrene combustion. Environ Sci Technol 32(15):2301–2307.

Swaine DJ (2000) Why trace elements are important. Fuel Process Technol 65–66: 21–23.

Swartz RC, Schults DW, Ozretich RJ, Lamberson JO, Cole FA, DeWitt TH, Redmond MS, Ferraro SP (1995) Σ PAH: a model to predict the toxicity of polynuclear aromatic hydrocarbon mixtures in field-collected sediments. Environ Toxicol Chem 14(11):1977–1987.

Teschke K, Hertzman C, Van Netten C, Lee E, Morrison B, Cornista A, Lau G, Hundal A (1989) Potential exposure of cooks to airborne mutagens and carcinogens. Environ Res 50(2):296–308.

Tornquist S (1985) Inverstigation of absorption, metabolism, kinetics and DNA-binding of intratracheally administered benzo[*a*]pyrene in the isolated, perfused rat lung: a comparative study between microcrystalline and particulate adsorbed benzo[*a*]pyrene. Chem Biol Interact 54:185–198.

USEPA (1997) Code of Federal Regulations, Title 40, Part 60, subparts D, Da, Db, Dc. Environmental Protection Agency, Washington, DC.

UNEP/GEMS (1991) Urban Air Pollution. Environment Library No. 4. United Nations Environmental Programme, New York.

Van de Wiel JAG, Fijneman PHS, Duijf CMP, Anzion RBM, Theuws JLG, Bos RP (1993) Excretion of benzo[*a*]pyrene and metabolites in urine and feces of rats: influence of route of administration, sex and long-term ethanol treatment. Toxicology 80(2–3):103–115.

Vayisoglu-Giray ES, Johnson BR, Fere B, Gizir AM, Bartle KD, Clifford AA (1998) Retention behaviour of polycyclic aromatic hydrocarbons during supercritical fluid chromatography with 1,1,1,2-tetrafluoroethane. Fuel 77(14):1533–1537.

Visser T, Martin S, Leonardus WJ, Jolanda W (1998) Identification of isomeric polycyclic aromatic hydrocarbons (PAH) in pyrolysates from ethynylated PAH by gas chromatography-Fourier infrared spectroscopy. Their relevance for the understanding of PAH rearrangement and interconversion processes during combustion. Fuel 77:913–920.

Wang X, Xue W, Zhu J, Gu Y, Sheng G, Fu J (1994) Characterization of polynuclear aromatic sulfur heterocycles in a coal extract by GC/MS. J Fuel Chem Technol 22:196–202 (in Chinese with English abstract).

Weyand EH, Bevan DR (1986) Benzo[*a*]pyrene disposition and metabolism in rats following intratracheal instillation. Cancer Res 46:5655–5661.

Wiersma DA, Roth RA (1983) Total body clearance of circulating benzo[*a*]pyrene in conscious rats: effect of pretreatment with 3-methylcholanthrene and the role of liver and lung. J Pharamacol Exp Ther 226:661–666.

Williams PT, Taylor DT (1993) Aromatization of tyre pyrolysis oil to yield polycyclic aromatic hydrocarbons. Fuel 72(11):1469–1474.

Williams PT, Keith DB, Gordon EA (1986) The relation between polycyclic aromatic compounds in diesel fuels and exhaust particulates. Fuel 65(10):1150R–1158R.

Wilson NK, Kuhlman MR, Chuang JC, Mack GA, Howes JE (1989) A quiet sampler for the collection of semivolatile organic pollutants in indoor air. Environ Sci Technol 23(9):1112–1116.

Wise SA, Schantz MM, Benner BA, Hays MJ, Schiller SB (1995) Certification of polycyclic aromatic hydrocarbons in a marine sediment standard reference material. Anal Chem 67:1171–1178.

Withey JR (1991) Pharmacokintics and bioavailability of pyrene in the rat. J Toxicol Environ Health 32:442–447.

Withey JR, Shedden J, Law FCP, Abedini S (1992) Distribution to the fetus and major organs of the rat following inhalation exposure to pyrene. J Appl Toxicol 12:223–231.

Withey JR, Shedden J, Law FCP, Abedini S (1993) Distribution of benzo[*a*]pyrene in pregnant rats following inhalation exposure and a comparison with similar data obtained with pyrene. J Appl Toxicol 13(3):193–202.

World Health Organization (WHO) (1987) Polynuclear aromatic hydrocarbons (PAH). In: Air Quality Guidelines for Europe. Regional Office for Europe, Copenhagen, pp 105–117.

Wolff RK (1989) Effects of repeated inhalation exposures to 1-nitropyrene, benzo[*a*]pyrene, Ga_2O_3 particles, and SO_2 alone and in combinations on particle clearance, bronchoalveolar laveage fluid composition, and histopathology. J Toxicol Environ Health 27:123–138.

Wolfgang FR, Hildemann LM, Mazurek MA, Glen RC (1991) Sources of fine organic aerosol. 1. Charbroilers and meat cooking operations. Environ Sci Technol 25(6):1112–1125.

Wolfgang FR, Hildemann LM, Mazurek MA, Glen RC (1993) Sources of fine organic aerosol. 5. Natural gas home appliances. Environ Sci Technol 27(13): 2736–2744.

Wolfgang FR, Hildemann LM, Mazurek MA, Glen RC (1997) Sources of fine organic aerosol. 8. Boilers burning no. 2 distillate fuel oil. Environ Sci Technol 31(10):2731–2737.

Wolfgang FR, Hildemann LM, Mazurek MA, Glen RC (1998) Sources of fine organic aerosol. 9. Pine, oak, and synthetic log combustion in residential fireplaces. Environ Sci Technol 32(1):13–22.

Xu XC, Chen CH, Qi HY (2000) Development of coal combustion pollution control for SO_2 and NOx in China. Fuel Process Technol 62:153–160.

Xue J, Liu GJ, Niu ZY, Zhang HY (2007) The impact factors of polycyclic aromatic hydrocarbons extracted from raw coal. Energy Fuels. 21(2):881–890.

Xue J, Liu GJ, Zhang HY, Niu ZY (2006) Research on the impact of PAHs extracted from raw coal in different times. Res Environ Sci 19(5):107–112.

Yelena YN, Steven JE, Barbara JT, Clifford PW, et al. (2002) Polycyclic aromatic hydrocarbons in the indoor and outdoor air of three cities in the U.S. Environ Sci Technol 36(12):2552–2559.

Yu Y, Gharaibeh A, Hawthorne SB, David JM (1995) Combined temperature/modifier effects on supercritical CO_2 extraction efficiencies of polycyclic aromatic hydrocarbons from environmental samples. Anal Chem 67:641–646.

Yue M, Gu X, Zou H, Zhu R, Su W (2003) Killer of health polycyclic aromatic hydrocarbons. J Capital Normal Uni 24(3):40–44 (in Chinese).

Zander M (1980) Polycyclic aromatic and heteroaromatic hydrocarbons. Handb Environ Chem 3(A):190–131.

Zhang HY, Liu GJ, Xue J (2005a) Study of polycyclic aromatic hydrocarbons (PAHs) and its environmental impact in coal and coal combustion products. J China Coal soc 30(suppl):97–101.
Zhang HY, Liu GJ, Xue J (2005b) The impact on the species, concentration and distribution of PAHs extracted by different solvents from raw coal. Environ Chem 5(24):613–616.
Zhong T, Yang W (2000) Efficient and environment friendly use of coal. Fuel Process Technol 62:137–141.

Manuscript received July 8, accepted October 4, 2006.

Rev Environ Contam Toxicol 192:29–66

Pasture Soils Contaminated with Fertilizer-Derived Cadmium and Fluorine: Livestock Effects

Paripurnanda Loganathan, Mike J. Hedley, and Neville D. Grace

Contents

Communicated by David H. Whitacre.

P. Loganathan (✉)
Fertilizer and Lime Research Centre, Institute of Natural Resources, Massey University, Tennent Drive, Private Bag 11222, Palmerston North, New Zealand

M.J. Hedley
Fertilizer and Lime Research Centre, Institute of Natural Resources, Massey University, Tennent Drive, Private Bag 11222, Palmerston North, New Zealand

N.D. Grace
AgResearch Limited, Grasslands Research Centre, Private Bag 11008, Tennent Drive, Palmerston North, New Zealand

I. Introduction

Pasture-based livestock production is a major system of agriculture in many parts of the world. Approximately one-fifth of the world's land area is used for livestock production, twice the area used for growing crops. Grazing livestock occupy approximately 3.5 billion ha of permanent pasture (FAO Statistics 2006). In the more intensively stocked pastures, pasture production has been increased through introduction of improved pasture species and regular application of fertilizers. In ryegrass and clover-based pastures, common in temperate climates, the main nutrients applied in fertilizers have been phosphorus (P) and sulfur (S). Nitrogen, sulfur, and potassium fertilizers are relatively free of contaminants, but P fertilizers, e.g., single and triple superphosphate (SSP, TSP), and diammonium phosphate (DAP) often contain many contaminants derived from the phosphate rocks (PR) used in their manufacture. If not managed appropriately, these fertilizer contaminants may cause adverse effects on livestock health, food quality, and environment.

Sauerbeck (1992) listed 13 potentially toxic elements that have significantly higher concentrations in PR than in the earth's crust. Considering the balance between current rates of P fertilizer application to agricultural land and rate of removal in crops, he concluded that arsenic (As), cadmium (Cd), fluorine (F), and uranium (U) concentrations in topsoils would continue to increase. McLaughlin et al. (1996a) conducted a similar assessment for the contaminants As, Cd, mercury (Hg), lead (Pb), and F associated with P fertilizers applied to dryland wheat and irrigated potato in southern Australia (Table 1). They concluded that the highest risk to the environment was asso-

Table 1. Estimated time taken for doubling the concentrations of contaminating elements in topsoils (0–100 mm depth) when P fertilizers are applied to wheat and potatoes in southern Australia.

Crop	Element	Input (g/ha)	Crop harvest (g/ha)	Net rate of addition (g/ha/yr)	Background soil concentration (mg/kg)	Years to double soil concentration
Wheat	As	1.0	0.30	0.70	4.0	7,500
	Cd	6.0	0.12	5.88	0.2	45
	Hg	0.1	0.03	0.07	<0.1	1,850
	Pb	4.0	0.30	3.70	21.0	7,500
	F	4,000	3.0	3,997	300	100
Potato	As	4.0	0.20	3.80	4.0	1,370
	Cd	20.0	2.50	17.50	0.2	15
	Hg	0.4	0.25	0.15	<0.1	870
	Pb	16.0	1.00	15.00	21.0	1,800
	F	16,000	10.0	15,990	300	25

Source: McLaughlin et al. (1996a).

Table 2. Estimated time taken for doubling the concentrations of Cd and F in topsoils (0–100 mm depth) when P fertilizers are annually applied to grazed pastures in New Zealand.

Element	Input[a] (g/ha)	Pasture uptake[b] (g/ha)	Animal intake[c] (g/ha)	Animal removal[d] (g/ha)	Net rate of addition[e] (g/ha/yr)	Background soil concentration (mg/kg)	Years to double soil concentration[f]
Cd	8.4	2.0	1.6	0.0016	8.3984	0.3	36
F	6,000	50	40	20	5,880	300	51

[a]30 kg P/ha/yr, New Zealand Fertilizer Manufacturers' Research Association Limit of 280 mg Cd/kg P, 200 g F/kg P.

[b]10,000 kg dry matter/ha/yr, 0.2 mg Cd/kg dry matter, 5 mg F/kg dry matter.

[c]80% utilization of pasture by animals.

[d]0.1% Cd and 50%F permanent removal.

[e]Column 2 minus column 5.

[f]Bulk density of 1 g/cm^3.

ciated with Cd and F because the time taken to double their concentrations in topsoils as a result of P fertilizer application was less than (<100 yr) that for the other elements (>1000 yr). Similar calculations carried out for New Zealand pastoral soils for Cd and F showed the same time scale for doubling the concentrations of these elements in the topsoils (Table 2).

This chapter reviews the research carried out on the effect of Cd and F on grazing livestock, the sources of Cd and F inputs to grazed pastures, their reactions in soils, fertilizer and grazing management strategies to control their accumulation in soils, plants and grazing livestock, and possible future research programs.

II. Cadmium Accumulation in Edible Offal Products and Fluorine Toxicity in Livestock

Cadmium is not essential for humans, animals, and plants. However, excessive intake of Cd by humans and animals can cause health risk as a result of renal tubular dysfunction and other diseases. Fluorine has some essential role in humans and animals but not in plants. However, excessive intake of F can cause fluorosis in humans and animals.

As a result of Cd accumulation in soils from P fertilizer use, there have been reported also many incidences of elevated Cd concentration in the edible offal products of grazing livestock in New Zealand (Loganathan et al. 2003), Australia (Koh et al. 1998), and other countries (Prankel et al. 2005). For example, during 1988–1992, one in every five sheep or cattle kidneys tested in New Zealand had a Cd concentration above the maximum permissible concentration (MPC) of 1 mg/kg fresh weight (Marshall 1993; Roberts et al. 1994), an acceptable value set by the New Zealand Department of Health until the end of 2002 (Table 3). Cadmium concentrations

Table 3. Maximum permitted concentrations of cadmium in food in New Zealand.

	Cadmium Content (mg/kg, fresh weight)	
	Before Dec. 2002	Since Dec 2002[a]
Kidneys	1	2.50
Liver	1	1.25
Meat flesh (muscle)	1	0.05
Leafy vegetables	1	0.10
Root and tuber vegetables	1	0.10
Wheat	1	0.10

[a]Australian and New Zealand Food Authority (ANZFA).
Source: Furness (2001).

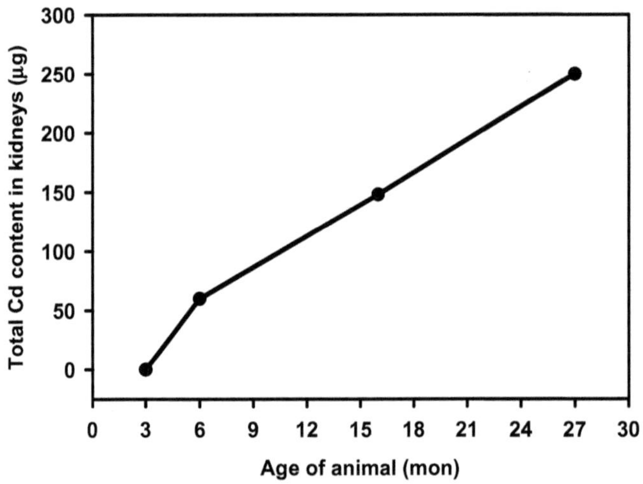

Fig. 1. Total accumulation of Cd in kidney of grazing sheep (Lee et al. 1996. *http://www.publish.csiro.au/nid/43/issue/1061.htm*. Reproduced with permission from CSIRO Publishing and *AJAR*.)

in kidneys increase with the age of the animal (Fig. 1). Therefore, to meet the European Communities import requirements, kidneys from older animals [>2.5 yr in New Zealand (Roberts et al. 1994) and >4 yr in Australia (McLaughlin et al. 2000)] were excluded from shipments for human consumption. Since December 2002, new standards for Cd in foods have been adopted by the Australian and New Zealand Food Authority (ANZFA). For edible offal products, these are higher than the previous New Zealand standards, but are 20–50 times higher than those for muscle (see Table 3).

The varying standards for animal products reflect the natural protective mechanism operating in animals whereby the blood Cd is sequestered by metallothionein and then is removed from circulation in the kidneys and liver (Petering and Fowler 1986). The percentage of slaughtered animals having kidney Cd concentration exceeding the new MPC in New Zealand is likely to be less than 3% based on the earlier survey results (Marshall 1993). However, some foods, for example, wheat, may exceed the new permissible levels.

Until recently, most of the concern with F toxicity had been associated with cattle and sheep deaths caused by acute F poisoning after ingesting freshly applied single superphosphate (SSP) fertilizer or basic slag (Clark et al. 1976; Cronin et al. 2000; Jones and Jones 1962; O'Hara and Cordes 1982). For example, O'Hara et al. (1982) noticed acute fluorosis in sheep with a short-term (few days) high F intake of >3000 mg/kg dry matter (DM) diet (lethal dose, 50–100 mg F/kg live weight), resulting in kidney failure and death. However, chronic F toxicity in sheep and cattle can be induced by much lower dietary F concentrations of 30–170 mg/kg DM, over months or years, causing damage to teeth, jaw, and bones (Clark and Stewart 1983).

III. Cadmium and Fluorine Sources

A. Cadmium

There are four major sources of Cd input in agricultural soils. One source is the geogenic Cd derived from parent rock (<0.03–2.6 mg/kg), with igneous rocks at the low end and sedimentary rocks such as black shale at the high end of the concentration range, respectively (Adriano 2001; Cook and Freney 1988). The other three sources are anthropogenic, namely, the use of phosphate fertilizer, atmospheric deposition arising from industry, and sewage sludge and industrial waste application to agricultural lands (Fig. 2). These three anthropogenic sources are major contributors to the Cd burden in agricultural soils in Europe and some parts of North America. However, in countries such as New Zealand and Australia where agriculture is the main industry, phosphate fertilizers are the major source of Cd. Atmospheric deposition and sewage sludges are important only in certain localities. For example, Fergusson and Stewart (1992) reported that within a 20-km radius around Christchurch, New Zealand, the Cd input to land from atmosphere was 1–2 g Cd/ha/yr, equivalent to 10%–20% of Cd input from fertilizers. But, at 30–80 km from the city the Cd addition from the atmospheric deposition was 0.02–0.05 g/ha/yr, or <1% of Cd addition in fertilizers. In contrast, in Europe and the United States, the amount of Cd added to soils from the atmosphere in rural areas varied from 1 to 25 g/ha/yr (Jackson and Alloway 1992); in many areas, approximately equal inputs of Cd from atmosphere and fertilizers were reported (Louekari 1996; Nicholson et al. 1996).

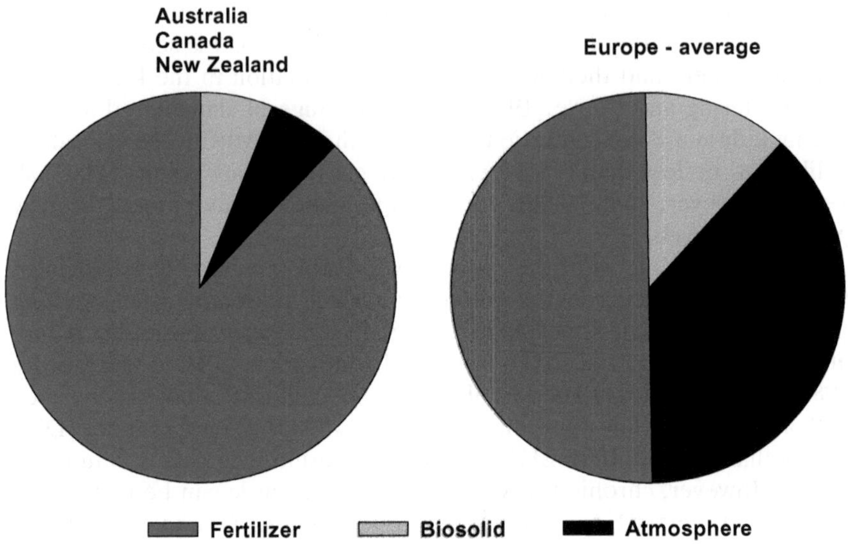

Fig. 2. Diagrammatic representation of relative contribution of Cd from fertilizer, atmospheric, and biosolid sources to agricultural land in various countries. (Redrawn from McLaughlin 2000.)

Soil parent material is unlikely to have significant influence on topsoil Cd concentration in fertilized pastoral farms because the amounts of Cd added in fertilizers are much greater than the contribution from the parent material. Influence of parent material was evident only in some unfertilized native soils (Roberts et al. 1994). Zanders (1998) studied the distribution of Cd throughout the soil profiles of 17 soil types on a farm in New Zealand that had a long P fertilizer history and reported that Cd concentration generally decreased with depth and that soil parent material contributed little to the topsoil Cd concentration. In most of the profiles, insignificant levels of Cd were detected in the C and lower B horizons.

B. Fluorine

Fluorine is a common constituent of all rocks. It forms 0.078% of the earth's crust (Simons 1950), and parent rocks of soils generally contain between 180 and 1000 mg F/kg (Fleischer and Robinson 1963). As found for Cd, the soil parent material, the atmosphere, and P fertilizers are also important for F, with P fertilizer being the main source of F in pastoral topsoils. In addition, in certain countries (e.g., Iceland, New Zealand), ashfall from volcanic eruptions can also be an important source of F (Coote et al. 1997; Cronin et al. 2003; Weinstein and Davison 2004). As was observed for Cd, the F concentrations of fertilized topsoil seldom reflect the F concentrations of the soil parent material (Loganathan et al. 2003) because the soil environment is often influenced by anthropogenic sources.

IV. Cadmium and Fluorine Levels in Phosphatic Fertilizers

Common P fertilizers for pastoral soils, SSP, TSP, DAP, or partially acidulated phosphate rock (PAPR), are manufactured by acidulation of phosphate rocks (PR). Small amounts of the more reactive PRs (RPR) are directly applied as low-cost, slow-release P sources. These fertilizers contain varying concentration of Cd depending on the origin of the PR used to make the fertilizer (Table 4). The location of Cd within the PRs is in the

Table 4. Cd, F, and P concentrations of a range of phosphate rocks and fertilizers.

Phosphate fertilizer	Cd ($mg\,kg^{-1}$)	F (%)	P (%)
Kola PR	0.2		17.2
Chatham Rise PR	2	3.0	8.9
North Florida PR	3	4.0	13.3
Phalaborwa PR	4		17.2
Jordan PR	5	3.8	14.0
Egypt (Quseir) PR	8		12.7
Mexico PR	8	4.1	14.0
Makatea Island PR	10	3.2	13.0
Sechura PR	11	3.4	13.1
Arad PR	12	4.0	14.1
Khouribga PR[a]	15		13.8
Syria PR[a]	16		13.3
Algeria PR[a]	18		12.5
Nahal Zin PR[a]	31		13.3
Gafsa PR	38	4.1	13.4
Morocco (Boucraa) PR	38		15.7
Youssoufia PR[a]	40		14.2
Tunisia PR[b]	40		
North Carolina PR	41	3.5	13.0
Christmas Island PR	43	2.2	15.3
Togo PR[a]	60		15.9
Taiba PR[a]	75		15.9
Senegal PR[b]	87		
Ocean Island PR	99		
Nauru PR	100	3.0	15.6
SSP	5–30	1.08–1.84	9.0
TSP	70	1.3–2.4	21
DAP	7–75	1.2–3.0	20

PR, phosphate rock.

Data sourced from Davister (1996)[a] and Roberts and Stauffer (1996).[b] All other data from Loganathan et al. (2003).

apatite lattice substituting for Ca (Cook and Freney 1988), and within single superphosphate, Cd is associated with the Ca in phosphate and sulfate components (Williams and David 1973). The PRs of sedimentary origin, especially the guano-derived ones (Ocean and Nauru), have a very high concentration of Cd, and those of igneous rocks (Phalaborwa, Kola) have very low Cd concentrations. The low Cd concentration in igneous rocks is attributed to the volatilisation of Cd at 765°C, well below the temperature at which the rocks were formed.

The Cd concentration of manufactured P fertilizers (see Table 4) varies widely depending on the process and the PR used in the manufacture. For example, the Cd concentration of SSP was reported to be 5–30 mg/kg (Loganathan et al. 2003) and that of DAP was 7–75 mg/kg (McLaughlin et al. 1996a; Soler and Rovira 1996). Because the amount of Cd applied to land depends on both the Cd and P contents of the fetilizer, it is appropriate to express Cd concentration per unit P concentration in fertilizer. On this basis, it is often found that high-analysis P fertilizers such as triple superphosphate (TSP) give rise to less accumulation of Cd in soils.

The F concentration in PRs is generally about 3%–4%, i.e., 300–10,000 times higher than that of Cd concentration (see Table 4). Unlike Cd, the range of F concentrations found in different PRs, especially in fluroapatites, is relatively narrow. However, during the manufacture of phosphate fertilizers using PRs a significant quantity of the F is released as HF and SiF_4 gases, and the level of F in the fertilizers is likely to depend on the recovery of these gases in the scrubber and whether scrubber fluids are reblended with the fertilizer (Weinstein and Davison 2004). Typical SSP applications of 10–30 kg P/ha/yr are likely to add at least 1–5 kg F/ha/yr to pasture soils.

V. Effect of Phosphatic Fertilizers on Cadmium and Fluorine Accumulation in Pastoral Soils

Numerous studies conducted in many countries have shown that long-term use of P fertilizers has increased Cd and F concentrations in the topsoils (Cronin et al. 2000; McLaughlin et al. 1996a). In New Zealand, a national survey (Roberts et al. 1994) of Cd in pasture and nonagricultural topsoils (0–75 mm depth) showed that in pasture topsoils ($n = 312$) that had been receiving regular P fertilizer inputs, the Cd concentration ranged from 0.04 to 1.53 mg Cd/kg, with a mean value of 0.44 mg Cd/kg. In comparison, the nonagricultural soils ($n = 86$) receiving no fertilizer had Cd concentration in the range 0.02–0.77 mg Cd/kg, with a mean value of 0.20 mg Cd/kg (Fig. 3).

In a study on irrigated pasture grazed with sheep in New Zealand that was annually topdressed with different rates of SSP for 44 years, Gray et al. (1999a) noticed that the topsoil Cd (0–75 mm) concentration increased from 0.05 to 0.45 mg Cd/kg soil as a result of SSP application (376 kg SSP/ha/yr) while the rate of Cd accumulation in the topsoil was 7.8 g Cd/ha/yr (Fig. 4).

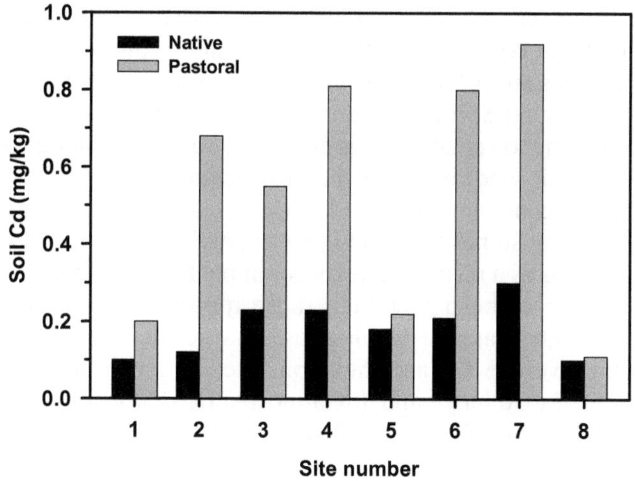

Fig. 3. Mean soil Cd concentrations (0–25 mm depth) for native and pastoral sites by soil group: Site 1, Recent Soils; Site 2, Granular Soils; Site 3, Glay Soils; Site 4, Organic Soil; Site 5, Brown Soils; Site 6, Allophanic Soils; Site 7, Pumice Soils; Site 8, Pallic Soils. For soil classification, see Hewitt (1998). (Redrawn from Roberts et al. 1994. Courtesy of Royal Society New Zealand.)

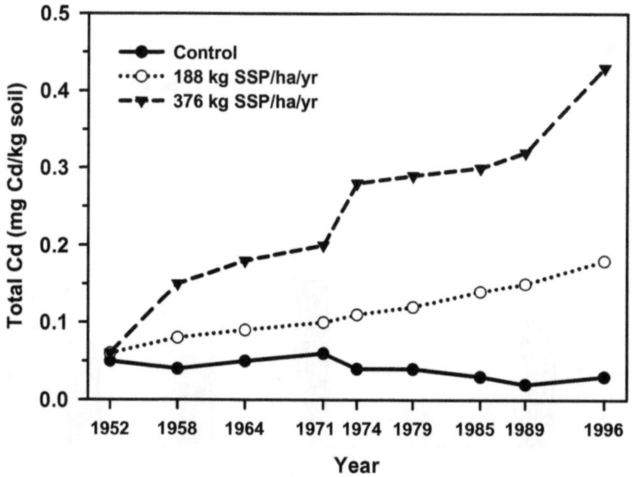

Fig. 4. Total Cd concentrations in topsoils (0–75 mm) for different rates of fertilizer treatments (Winchmore, New Zealand). (Gray et al. (1999a) The effect of long-term phosphate fertiliser application on the amounts and forms of cadmium in soils under pasture in New Zealand, vol 54, pages 267–277, Fig. 1. With kind permission of Springer Science and Business Media.)

In another New Zealand study on hill country pasture grazed with sheep, Loganathan et al. (1995) reported Cd accumulation rates of 11.3 g Cd/ha/yr for the high slopes and 17.8 g Cd/ha/yr for the low slopes of the hill pastures 20 years after annual SSP application at the rate of 425 kg SSP/ha/yr. The larger value for the lower slopes was attributed to greater amounts of fertilizer input per unit land area and the greater amounts of Cd returned in animal excreta because animals spend proportionately more time resting and defecating on lower slopes than on higher slopes. The rates of Cd accumulation in the topsoils as a result of P fertilizer application in New Zealand pastoral soils appear to be higher than those reported elsewhere, probably because of the higher rate of P fertilizer use on volcanic soils. For example, in Denmark, the average Cd accumulation rate for fertilizer Cd input was reported to be 3 g Cd/ha/yr and for atmospheric deposition the rate was 2 g Cd/ha/yr (Louekari 1996). In a 70-yr permanent grass experiment at the Rothamsted Experimental Station, UK, Nicholson et al. (1996) reported that the average Cd accumulation rate in the top 225 mm soil was 5.8 g Cd/ha/yr whereas the contribution from P fertilizer input to this rate was 3.3 g Cd/ha/yr. The continuous use of phosphate fertilizers has also been shown to significantly increase topsoil F concentration in pastoral soils (Fig. 5).

Total Cd and F concentrations in regularly fertilized topsoils are often positively correlated with total P concentrations (Figs. 6, 7). Agricultural

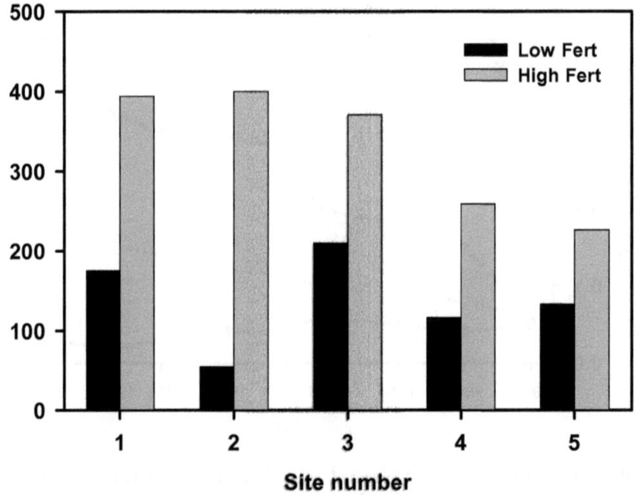

Fig. 5. Mean F concentrations (0–75 mm depth) for native/pastoral sites topdressed with low rates of, or no, P versus pastoral sites topdressed with high annual P rates. Site 1, Limestone Downs, Allophanic Soil; Site 2, Limestone Downs, Ultic Soil; Site 3, Normanby, Allophanic Soil; Site 4, Ballantrae, Pallic/Brown Soil; Site 5, Palmerston North, Pallic Soil. For soil classification, see Hewitt (1998). (Redrawn from Loganathan et al. 2001, 2003.)

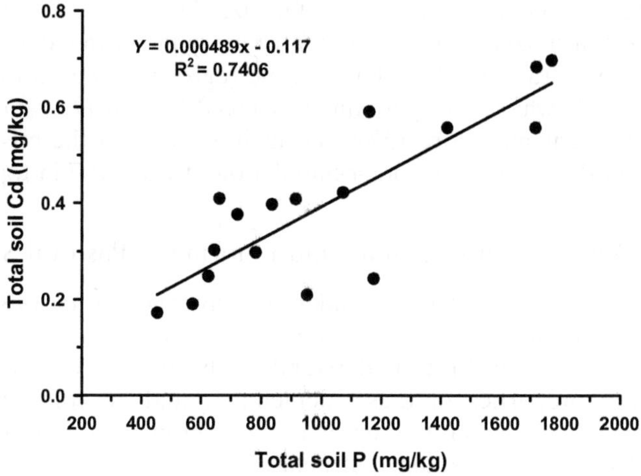

Fig. 6. Relationship between total P and total Cd in the 0–30 mm soil depth of high fertility farmlets at Ballantrae, New Zealand. (Loganathan P, Mackay AD, Lee J, Hedley MJ (1995) Cadmium distribution in hill pastures as influenced by 20 years of phosphate fertilizer application and sheep grazing. Aust J Soil Res 33:859–871. *http://www.publish.au/nid/43/issue/235.htm*. Reproduced with permission from CSIRO Publishing *AJSR*.)

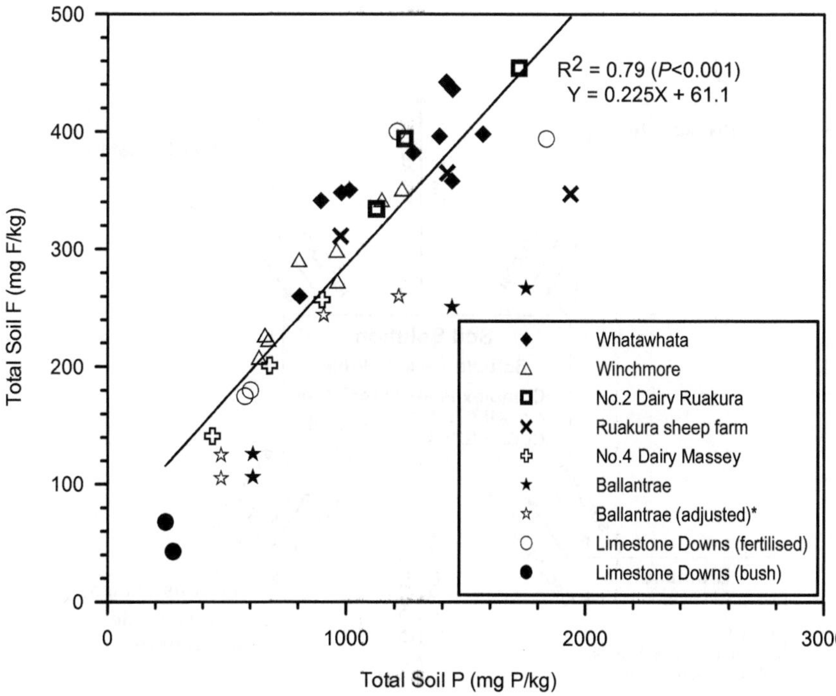

Fig. 7. Relationship between total P and total F in the 0–75 mm soil depth in New Zealand pastoral soils. (Reprinted from Loganathan P, Hedley MJ, Wallace GC, Roberts AHC (2001) Fluoride accumulation in pasture forages and soils following long-term applications of phosphorus fertilizers. Environ Pollut 115:275–282. With permission from Elsevier).

soil P concentrations are a reflection of fertilizer history; therefore, it would appear that P fertilizers are a major contributory factor in Cd and F accumulation in pastoral soils. The slopes of the regression lines in Figs. 6 and 7 are 490 and 5, which are approximately the P/Cd and P/F ratios in SSP used in New Zealand before 1996; this again shows that there is a strong link between P fertilizer use and accumulation of Cd and F in topsoils.

VI. Dynamics of Cadmium and Fluorine in Pastoral Soils

The Cd and F added in fertilizers undergo many reactions in soils depending on the nature and properties of the soil solid and solution phases and other environment (Fig. 8). To understand the dynamics of Cd and F in soils the chemical properties of these elements should be considered first. Cadmium is a heavy metal that exists as a cation in solution whereas F is a highly electronegative anion that behaves similar to P in soil solution. Both F and P are mostly specifically adsorbed to soil components. The common soil properties that influence the partitioning of Cd and F between soil solid and solution phases are listed in Table 5. Increases in Cd and F concentration in soil solution leads to increases in their mobility and bioavailability.

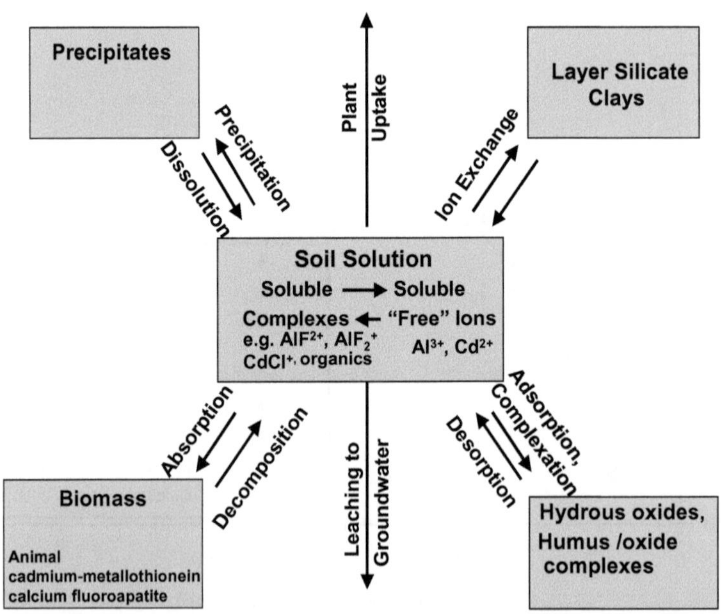

Fig. 8. Dynamics of Cd and F reactions in soil. (Adapted from Adriano 2001.)

Table 5. Soil properties influencing the dynamics of Cd and F in soils.

Soil property	Cd	F	References
pH	Increase of solution conc. due to decrease of pH (see text)	Decrease of pH below 5.5 and increase above 6.5 increase solution conc.	Boekhold and Van der Zee (1992); Gray et al. (1998); Larsen and Widdowson (1971); Loganathan et al. (2006); McBride et al. (1997); Merry (1988); Williams and David (1973)
Fe and Al oxides and allophane	Decrease of solution conc. due to sorption	Decrease of solution conc. due to sorption	Bower and Hatcher (1967); McLaughlin et al. (1996a); Omueti and Jones (1977); Pickering (1985)
Mn oxides	Decrease of solution conc. due to sorption	No significant influence	Pickering (1985); Zasoski and Burau (1988)
Layer silicates	Decrease of solution conc. due to sorption	Decrease of solution conc. due to sorption	Kau et al. (1998); McLaughlin et al. (1996a); Pickering (1985)
Organic matter	Decrease of solution conc. due to sorption (increase cation-exchange capacity, chelation)	Indirect effect (see text)	Boekhold and Van der Zee (1992); Degryse and Smolders (2006); Gray et al. (1998, 1999a,b); McBride et al. (1997); Loganathan et al. (2006); Pickering (1985)
Other ions in solution	Increase of Cl conc. increases solution conc. (weakly adsorbed CdCl complexes); Ca and Zn increase solution Cd conc. (competitive adsorption); P induced Cd adsorption	Increase of solution F conc. due to increase Al conc. High rates of Ca application reduce solution F conc.	Adriano (2001); Bolan et al. (1999); Fung and Wong (2004); Manoharan et al. (1996); McLaughlin et al. (1996a); McLaughlin et al. (2001)
Eh (redox potential)	Increase redox potential decrease solution conc. (CdS formation); Fe/Mn oxide dissolution causing increase solution Cd conc.	Increased redox potential likely to reduce F^- sorption but no published data	Adriano (2001); Brown et al. (1989)
CEC	Increased cation-exchange capacity, decreased solution conc.	—	Bolan et al. (2005)

conc., concentration.

Soil pH has often been found to be the most important single soil property that determines the Cd and F mobility and bioavilability in soils (Adriano 2001; Sauerbeck 1992). Unlike other heavy metals, such as Cr, Ni, Co, Cu, and Pb, Cd and Zn can be mobilized even in slightly acidic conditions (Sauerbeck 1992).

Gray et al. (1999b) investigated the Cd mobility characteristics of 29 topsoils from predominantly pastoral sites belonging to 10 soil orders in New Zealand and concluded that pH was the most dominant soil variable affecting solution Cd concentration and sorption and desorption of native and added Cd in these soils. Similarly, in the Netherlands, at sites from various land uses, including grassland, and different soil types it was found that solubility of Cd was controlled mostly by soil pH and, to a much lesser extent, by dissolved organic carbon (Castilho et al. 1996; Römkens and Vries 1995). Bolan et al. (2005) reported three possible reasons for the increase in solution Cd concentration with decrease in soil pH. First, in variable-charge soils, a decrease in pH causes a decrease in surface negative charge, thereby lowering Cd adsorption; second, a decrease in soil pH is likely to decrease the proportion of Cd present as the hydroxy species of Cd ($CdOH^+$), which are adsorbed more strongly than the simple Cd^{2+} ion; and third, acidification causes the dissolution of Cd salts, increasing the concentration of Cd in soil solution. The extent of the increase, however, is dependent on the soil type. For instance, the effect of pH is much less pronounced in Alfisols or net permanently charged soils than in strongly weathered or variable charge soils (Naidu et al. 1994).

The common management practice used to reduce soil acidity is application of lime, also recommended by many to reduce Cd uptake by plants (Loganathan et al. 2003; Louekari 1996; McLaughlin et al. 1996a). Glasshouse and laboratory liming trials showed that liming reduced Cd uptake by plants (McLaughlin et al. 1996b; Williams and David 1973). However, consistent results for the effects of lime have not been obtained under field conditions. For example, Öborn and Jansson (1998) reported that of the seven long-term field trials conducted in Sweden on the effect of lime on Cd uptake by spring wheat and potatoes, lime application increased plant Cd levels in three trials but decreased levels in three other trials. Similar inconsistent results were reported for Australian soils by McLaughlin et al. (1996b). Possible reasons for lime application increasing plant Cd uptake are lime increases root growth by supplying Ca and ameliorating soil acidity, which results in increasing access and absorption of Cd; Ca in lime desorbs the Cd in soil, thus increasing soil solution Cd concentration; lime decreases Zn availability in soil, thereby reducing Zn competition with Cd for plant uptake; and Ca supply from lime increases root cell membrane permeability, thus helping roots to take up more Cd from soil.

Fluorine solubility is lowest between 5.5 and 6.5 and increases dramatically at lower and higher pH (Larsen and Widdowson 1971; Omueti and Jones 1977) (Fig. 9). Increased solubility at low pH is caused by F-forming

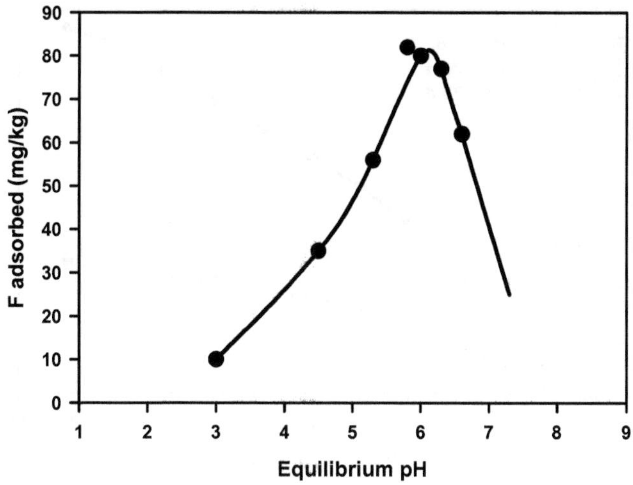

Fig. 9. Effect of pH on F^- adsorption by an A horizon of Cisne soil (Mollic Albaqualf) from Illinois (USA). Initial F concentration was 4 mg/L. (Redrawn from Omueti and Jones 1977.)

complexes with Al in the soil solution, leaving few free F (F^-) ions in soil solution for sorption (Manoharan et al. 2007). At high pH, the increasingly negative electrostatic potential decreases the retention of F on the soil. Increases in F^- concentration in soil solution also result from the displacement of adsorbed F^- by increased concentration of OH^- in soil solution at higher pH (Larson and Widdowson 1971).

VII. Cadmium and Fluorine Forms in Pastoral Soils

Gray et al. (1999a) chemically fractionated the Cd in several pastoral soils enriched with fertilizer-derived Cd and found that Cd was largely associated with an organic and a residual fraction. The lowest Cd concentration was in an exchangeable fraction. Although in the case of F the greatest proportion of the F in pastoral soils is associated with the Fe and Al oxides and residual fraction, the proportion in the organic fraction appears to be very low (Loganathan et al., unpublished data).

Cadmium availability decreases with increasing time after Cd application to soils. In a New Zealand pastoral soil it has been shown that the ability of sorbed Cd to desorb back into solution decreases with increasing contact time with the soil (Gray et al. 1998) (Fig. 10). Adriano (2001) reported that in soils where the Cd is present at background level the residual Cd fraction is usually dominant, whereas in most highly contaminated soils this fraction may not account for a large proportion of the total soil Cd. On the other hand, soil aging may also lead to increases in Cd

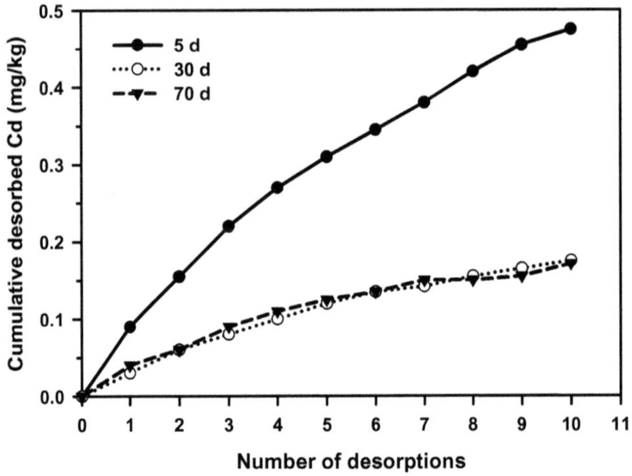

Fig. 10. Effect of contact period (5, 30, and 70 d incubation at field capacity moisture content) on cumulative desorption of added Cd from Waiareka clay loam (Ultisol) in New Zealand. (From Gray et al. 1998. *http://www.publish.csiro.au/nid/85/issue/235.htm*. Reproduced with permission from CSIRO Publishing and *AJSR*.)

bioavailability if Cd is originally associated with organic matter that is degrading with time (McLaughlin et al. 2000).

The quantity of Cd and F in soil solution represents <1% of their total soil contents (Gray et al. 1999b; Loganathan et al. 2006; Taylor and Percival 2001). In soil solution, Cd and F occur both as free ionic forms and complexed with other ions (see Fig. 8). For example, Taylor and Percival (2001) noticed that 55%–90% of the total inorganic Cd in acidic [soil pH (water), 5.2–7.0] pastoral soil solutions exists as Cd^{2+} with the majority of the remaining Cd in the form of chloro- and sulfato-complexes of Cd. In Australian saline soils, up to 75% and 20% of the inorganic Cd in soil solution was present as chloride and sulfate complexes, respectively (McLaughlin et al. 1996a). In contrast to some other heavy metals such as Cu and Pb, organic ligands do not significantly complex Cd (Adriano 2001) (the order of the strength of complexation is Cu > Pb > Cd). The complexation of Cd with dissolved organic matter is weak because of the competition for binding sites with Ca, which is generally present in much higher concentration in soil solution. The concentrations of Cd species in some Australian saline/sodic soils having dissolved organic carbon concentrations of 78–784 mg/L were only slightly altered when speciation was calculated using GEOCHEM-PC (Parker et al. 1994) with and without inclusion of dissolved organic carbon in the calculation (McLaughlin et al. 1997b).

Chien et al. (2003) compared Cd uptake by upland rice and soybean from soils treated with granulated and compacted PK fertilizer (SSP/KCl) as against bulk-blend PK fertilizer (both treatments had 1:1 weight ratio of P:K) in a pot trial and found that the Cd concentration in plant tissues was higher for the granulated PK fertilizer treatment than for the bulk-blend fertilizer treatment but the crop yields were same for the two treatments. They explained that the enhanced Cd uptake by plants in the granulated fertilizer treatment was probably the result of the formation of soluble CdCl complexes as a result of close contact between Cl and Cd in the granulated fertilizer.

In acid soils, F in soil solution is mainly complexed to Al, and the proportion of F complexed with Al increases with an increase in acidity mainly because of increasing soil solution Al concentration. In soil solutions extracted by 0.01 M $CaCl_2$ from highly acidic (pH in 0.01 M $CaCl_2$, 4.7–4.8) Australian pastoral topsoils (0–100 mm depth), 95%–98% of the F was estimated to be complexed with Al (McLaughlin et al. 2001). However, in the subsurface soils (200–300 mm depth) that had pH (in 0.01 M $CaCl_2$), >5.2, F^- was the dominant form of F in soil solutions. Similarly, in slightly acidic (pH in water, 5.4–5.9) New Zealand pastoral soils, only up to 55% of the F in soil solutions was estimated to be complexed to Al and the remaining F in soil solution was in the uncomplexed F^- form (Manoharan et al. 1996).

VIII. Soil Profile Redistribution of Cadmium and Fluorine Associated with Fertilizers

Cadmium and F associated with P fertilizers do not appreciably move down the soil profile, especially in pastoral soils that are rich in organic matter and Fe and Al oxides. Comparing the downward movement of Cd, F, and P following 10 yr annual application of SSP (30 and 60 kg P/ha/yr) in a pastoral soil, it was observed that the relative mobility of these three elements was similar (Fig. 11). More than 50% of the total quantity of the elements accumulated in the profile remained within the top 75 mm soil. Similarly, in six other New Zealand pastoral soils with long-term P fertilizer application histories, 33%–66% of fertilizer-derived F was recovered in the top 75 mm soil (Loganathan et al. 2001). The low rate of downward movement of F and P is caused by their relatively strong adsorption onto clay minerals and oxides of Fe and Al (Bower and Hatcher 1967), and that of Cd is caused by its adsorption onto clay minerals and oxides of Fe and Al as well as to organic matter (Adriano 2001). Because of the strong retention of Cd and F in surface soil layers, the leaching potential of these elements to contaminate groundwater is expected to be low; however, the strong retention may lead to increased intake of these elements by grazing livestock if they ingest large amounts of soil (see later section).

In a range of New Zealand pasture soils ($pH_{(water)}$ 5.5–6.0 at 0–50 mm depth) that were regularly fertilized with P, Gray et al. (2003) reported that the amount of Cd lost through leaching below 250 mm depth represents

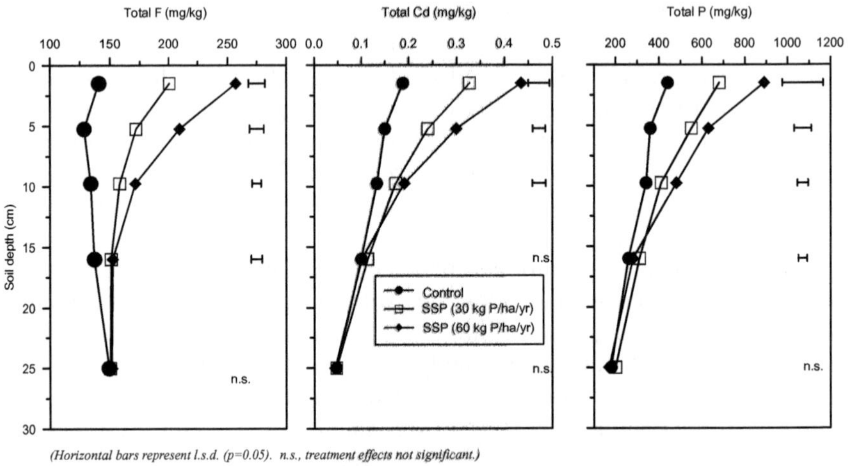

Fig. 11. Effect of 10yr annual applications of Single Superphosphate (SSP) to sheep-grazed pasture on the downward movement of F, Cd, and P in a Pallic Soil (an Aeric Fragiaqualf in US Soil Taxonomy). (Reprinted from Loganathan P, Hedley MJ, Wallace GC, Roberts AHC (2001) Fluoride accumulation in pasture forages and soils following long-term applications of phosphorus fertilizers. Environ Pollut 115:275–282. With permission from Elsevier).

only between 0.07% and 0.23% of the total Cd (0.19–0.69 mg Cd/kg for 0–250 mm depth) in the soils. The Cd concentrations in drainage waters in their study were substantially less than the maximum acceptable drinking water standard of 3 μg/L set by the New Zealand Ministry of Health. However, in highly acidic soils, high rates of Cd and F addition can lead to groundwater contamination, especially for coarse-textured soils with shallow water tables. For example, Degryse and Smolders (2006) reported significant quantities of Cd leached below 700 mm depth from several acid [pH (0.01 M $CaCl_2$) 3.2–4.5] sandy Spodosols in Belgium polluted by Cd deposition from a zinc smelter. The mean soil solution Cd concentration at 700 mm in polluted soils (total Cd at 0–100 mm soil depth was 0.82 mg Cd/kg) was 15 μg Cd/L compared to 1 μg Cd/L for unpolluted soils (total Cd at 0–100 mm soil depth was 0.18 mg Cd/kg).

IX. Cadmium and Fluorine Accumulation in Pasture

Cadmium accumulation in pasture depends on Cd availability in soil and the type of pasture species. It is controlled by the total soil Cd content in topsoil and the various soil properties influencing its solubility (see Table 5). The composition of well-fertilized pastures is generally dominated by cultivars of grass and clover, although the pastures also contain some weed species, depending on the fertility of the soils. In general, the weed species accumulate more Cd than grasses and legumes (Roberts et al. 1994).

Capeweed (*Arctotheca calendula* L.) is a common component of pastures in southeastern and southwestern Australia (McLaughlin 2000), and the Cd concentration in this weed was reported to be 1.57 mg/kg compared to 0.1–0.3 mg/kg for most grasses and clover species (Merry 1988).

In New Zealand hill country pasture, land slope has opposite effects on soil and herbage Cd concentrations. Although the soil Cd concentration decreases with slope, pasture Cd concentration increases with slope (Loganathan et al. 1995; Roberts and Longhurst 2002). The lower soil Cd concentration in higher slopes results from less fertilizer input per unit land area when fertilizer is topdressed and transfer of Cd by animal grazing and excretion patterns (animals graze higher slopes, removing Cd, and add excreta Cd where they rest in lower slopes). The higher pasture Cd concentration in higher slopes results from a number of factors, including pasture species composition (more weeds, fewer grasses, and more dead matter) and stage of growth (later stage of growth containing mature stems and flower heads as a result of lower grazing pressure).

There is a strong seasonal influence on the Cd concentration in pasture plants. In New Zealand, it has been observed that the Cd concentration in pasture herbage was highest during autumn and lowest during spring (Loganathan et al. 1996). These seasonal differences in Cd concentration are mainly the result of differences in pasture growth rate. During spring, growth rate is highest and Cd concentration is lowest due to a dilution effect, while during autumn, growth rate is lowest, resulting in higher Cd concentrations.

Phosphate fertilizer application generally increases topsoil F concentration but seldom increases pasture herbage F concentration (Loganathan et al. 2001). McLaughlin et al. (1996a) reported that, on a highly fertilized Australian soil, F concentrations in pasture herbage were not significantly greater than in pasture growing on adjacent unfertilized soil. The low plant availability of F in soils is consistent with low soil solution F concentrations because F is strongly bound to Fe and Al oxides surfaces, allophane, and layer silicates (see Fig. 8; see Table 5). In highly acidic soils with large inputs of F, however, the F concentration of pasture may be high. For example, in an Australian pasture soil of $pH_{(CaCl2)}$ of 4.6, it was reported that the mean pasture herbage F concentration (22 mg F/kg DM) in P-fertilized plots (annually topdressed with 125 kg SSP/ha for 36 yr) was significantly higher than that (11 mg F/kg DM) in the unfertilized plots (McLaughlin et al. 2001). Similarly, of 12 soils around aluminium smelters in Norway, in 11 soils having pH 4.1–5.8 and a water soluble F of 1–6 mg/kg, the F concentration in ryegrass shoots was 3–9 mg/kg, but in a highly F-contaminated soil having a pH of 4.9 and water-soluble F concentration of 33 mg/kg the F concentration in ryegrass shoots reached levels of 15–18 mg/kg (Arnesen 1997). These pasture F concentrations are, however, lower than the dietary F tolerance concentration limit of 30–150 mg/kg reported for grazing livestock (Cronin et al. 2000).

Application of F has been shown to ameliorate Al phytotoxicity in nutrient solutions (Cameron et al. 1986; MacLean et al. 1992), which was suggested to be the result of mainly a decrease in the concentrations of toxic Al (Al^{3+} and Al hydroxy species) as a result of complexation of F with Al. Stevens et al. (1997) have shown that Al–F complexes are not toxic at lower concentrations but are toxic at higher concentrations, and toxicity depends on the proportion and concentration of different types of Al-F species in nutrient solutions (AlF^{2+}, AlF_2^{1+}, AlF_3^0). The behavior of F in soils is more complex than in nutrient solutions because the addition of F to acid soils will dissolve soil solid-phase Al, whereby the soil solution Al concentration will increase (Moore and Richie 1988). Recently, Manoharan et al. (2006) reported that addition of high rates of F (80–160 mg/kg soil) to strongly acidic soils ($pH_{(water)}$ 4.3–4.6) reduced barley root growth by the toxicity of high concentrations of Al–F complexes (100–300 μM) formed in soil solution. They also stated that, at the current recommended rates of P fertilizer application to moderately acidic pastoral soils ($pH_{(water)}$ 5.5–6.0), it is unlikely that the concentrations of Al–F complexes (10–30 μM) will be sufficiently high to cause phytotoxicity. However, continuous input of F to soils through the application of high rates of P fertilizers, or F-emitting industrial processes and increased acidification of soils, may become a F risk to plants (Stevens et al. 1997) and grazing livestock (Arnesen 1997; Stevens et al. 2000) in the future.

X. Cadmium and Fluorine Intake and Distribution in Livestock

A. Cadmium

Grazing animals accumulate Cd in their organs, particularly the liver and kidneys, from herbage and incidental soil ingestion. The Cd intake by sheep through soil ingestion is low, making up approximately 14%–19% and <2% of total Cd intake during winter and summer months, respectively (Lee et al. 1996). The amount of soil ingested can increase severalfold during the wetter winter/early spring when more soil adheres to herbage and the grazing pressures on the pastures are high (Fig. 12).

More than 99% of Cd intake by sheep is returned to the soil in feces because daily net absorption is very small (Fig. 13). Cadmium accumulates mainly in kidneys and liver. Its concentration in sheep liver (0.2–1.1 mg/kg wet wt) and kidneys (0.2–6.5 mg/kg wet wt) increases with animal age (Petterson et al. 1991), but the rate of increase decreases with age (see Figs. 1, 13). At 6 mon of age, sheep retain about 0.1% of the Cd intake in the kidneys, with this figure decreasing to 0.04% or less with increasing age.

Cadmium in cow and goat milk (2–6 μg/kg) (Rayment 1988) and sheep and cattle muscle (10–30 μg/kg wet wt) are very low. Cd measured in carcass meat in New Zealand (20–30 μg/kg wet wt) (Roberts and Longhurst 2002)

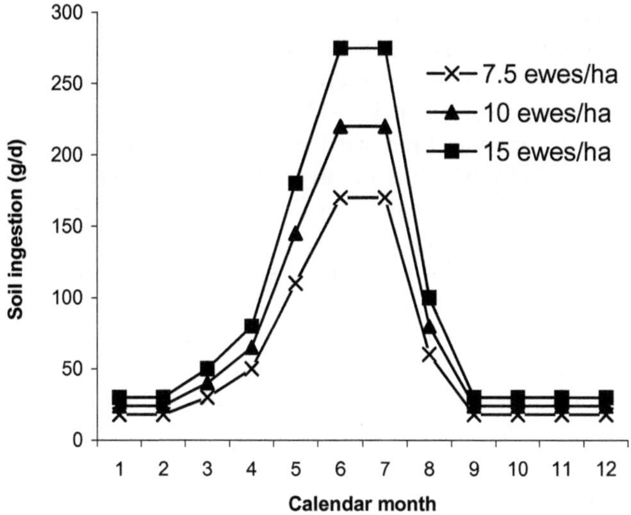

Fig. 12. Soil ingestion by grazing sheep as a function of calendar month and stocking rate in New Zealand: summer months, 12, 1, 2; autumn, 3, 4, 5; winter 6, 7, 8; spring 9, 10, 11. (Drawn from Healy 1968, 1973.)

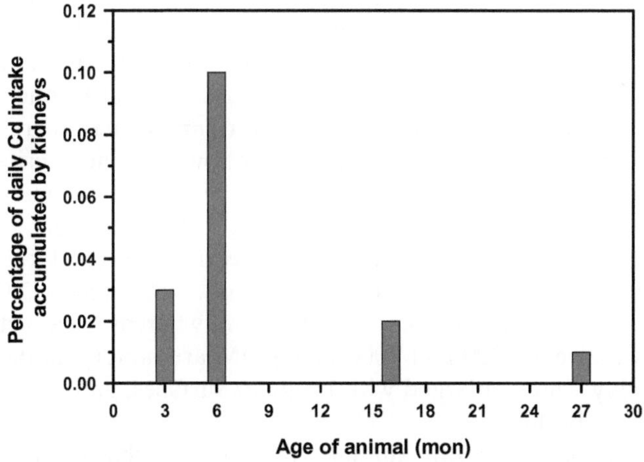

Fig. 13. Daily accumulation of Cd in kidneys of grazing sheep as a percentage of Cd intake. (From Lee et al. 1996. *http://www.publish.csiro.au/nid/43/issue/1061.htm*. Reproduced with permission from CSIRO Publishing and *AJAR*.)

and Australia (median, 10µg/kg wet wt) (Langlands et al. 1988) are significantly lower than the MPC of 50µg/kg fresh tissue in these countries (see Table 3). As the accumulation of Cd in kidneys is high whereas the amounts in milk and meat are low, the grazing animal can be viewed as a "filter" capable of ensuring that the amounts of Cd entering the food chain are low; this is similar to the case that when Cd-rich bran is removed from processed grains and roots are removed from edible parts in most plants, less Cd then enters the food chain (Sauerbeck 1992).

Cd and Zn in animals bind with the same metal-regulatory protein, metallothionein (MT), which is involved in transport and storage of both metals. Cd binds to MT more strongly than Zn, and it is this strong binding of Cd within the MT complex that provides the basis for Cd detoxification in animals (Petering and Fowler 1986). As Cd and Zn are mutually antagonistic, increasing dietary intake of Zn may be an effective method of reducing Cd absorption and, therefore, retention of Cd in animals. Phillips et al. (1997) showed that high dietary Zn reduced Cd absorption, resulting in greater excretion of Cd in feces. However, while Zn fed to sheep reduced the net accumulation of Cd in liver, it increased accumulation in kidney (Lee et al. 1991; Rounce et al. 1998). The accretion of Cd in the kidneys with high Zn doses has been associated with increased MT synthesis, which improved the efficiency of the kidney to sequester Cd (Bremner 1978; Lee et al. 1994; Peterson and Mercer 1988).

B. Fluorine

Grazing livestock accumulate F in bones and teeth as calcium fluoroapatites. Typical F concentration of enamel and dentine is 250–700 mg/kg and in bones is 250–900 mg/kg (Clark and Stewart 1983; Milhaud et al. 1992). Fluorine does not accumulate in the soft tissues. Typical F concentrations of sheep and cattle liver, kidney, thyroid, heart, and pancreas are <3 mg/kg dry wt. On the basis of cattle studies, the following bone F criteria have been established to describe fluorosis. Bone F concentrations of <2400, 2400–3200, 3200–4800, and 4800–7000 mg/kg DM reflect an innocuous, threshold, moderate, and severe fluorosis, respectively (Shupe 1980; Wheeler and Fell 1983). Bones with F concentrations >4800 mg/kg DM appear mottled but are normal on X-ray examination, whereas bone with F concentrations between 5000 and 7000 mg/kg DM are abnormal, being more porous, chalky white, enlarged with irregular surfaces, and having altered mechanical properties.

Fluorine intake by grazing animals via pasture herbage intake is much lower than through soil ingestion because of the very low F concentration in pasture herbage [generally <5 mg F/kg DM; Loganathan et al. (2001); McLaughlin et al. (1997a)] compared to topsoil F concentration [217–454 mg F/kg soil, Loganathan et al. (2001)].

Although the amount of soil ingestion is much less than pasture consumption, many factors including season and stocking rate may influence the magnitude of soil ingestion. The cooler, and wetter, winter/early spring months result in more soil adhering to leaves of plants, while higher stocking rates cause pastures to be grazed low, thereby increasing soil intakes

(Healy 1973). Thus, high F intakes associated with the soil ingestion occur only for 2–3 mon of the year.

Assuming a stocking rate of 2.9 dairy cows/ha and an annual herbage dry matter production of 15,000 kg/ha and using a herbage F concentration of 3 mg F/kg DM, the rate of F intake by dairy cows through pasture ingestion (80% dry matter utilization) is calculated to be <13 g F/cow/yr. This rate of total F intake is less than 0.7% of F added annually in fertilizers to soil, assuming a rate of fertilizer addition of 30 kg P/ha/yr, and F and P contents of fertilizer of 1.5% and 9% respectively. In comparison, the rate of annual F intake by a dairy cow through soil ingestion is 87–181 g/cow/yr, assuming an annual average daily soil ingestion rate of 1095 g/cow/d (Cronin et al. 2000) and soil F concentration of 217 to 454 mg F/kg soil (Loganathan et al. 2001). This annual rate of total F intake is equivalent to 1.8%–3.6% of F applied in P fertilizers/yr. Therefore, compared with the annual pasture F intake, the soil F intake is 3–6 times greater. Similar calculations for Cd intake by a dairy cow would show that Cd intake by pasture ingestion is 0.83 g/cow/yr, assuming 0.2 mg Cd/kg DM, while annual Cd intake by soil ingestion would be 0.12 g/cow, assuming 0.3 mg Cd/kg soil. Thus F ingestion by grazing animals is influenced more by soil ingestion, whereas Cd ingestion is influenced more by pasture consumption.

Weinstein and Davison (2004) reported that the proportion of ingested F absorbed and retained by an animal depends on the acidity of the digestive system, concentrations of chemical species such as Ca and Al in the digestive system, and the presence of calcified tissues in the body. They illustrated this by citing several examples of F absorption in humans, cattle, rabbit, voles, and insects. The values reported for F absorption and retention in sheep and cattle feeding on different diets are presented in Table 6. The values are lower for ingestion of soil compared to those for pasture and the soluble salt, NaF. The difference in the absorption values reported for soils probably results from the difference in the chemical form of F in the soils, which is dictated by soil properties such as pH, mineralogy, texture, and the period allowed for F added in fertilizers to react with the soils. It could also be caused by differences in the design and conduct of the trials.

The F absorbed by the animals is sequestered into bone or excreted in the urine. For sheep, urinary losses are 10%–23% of the ingested F, leaving 27%–40% of the absorbed F to be retained in the bone (Grace et al. 2003). Increased and decreased F intakes by sheep are immediately reflected by an increase and decrease in blood F concentrations, while increased F intake resulted in a much slower rate of increase of bone F concentrations. High bone F concentrations are indicative of animals being on a high F intake for an extended period (Table 7).

An indoor sheep feeding trial was conducted in New Zealand to simulate high soil ingestion during one winter season. In this trial, young sheep (14 mon old) were fed daily with lucerne (7 mg F/kg dry matter) or lucerne plus soil (184 mg F/kg dry matter) daily for 63 d. The blood F concentration

Table 6. Apparent absorption (% of intake) and retention (% of intake) of F by livestock.

Animal	F source	Trial design	Absorption	Retention	Reference
Sheep (14 mon old)	Lucerne	63-d feeding/8-d F balance trial	75–77	34–35	Grace et al. (2003)
	NaF + lucerne		69	27–31	
	Soil (10% by wt in lucerne feed, 600 and 1,751 mg F/kg soil) + lucerne		44–53	28–42	
Dairy cattle (>7 yr old)	Soil (10% by wt in lucerne feed, 1,452 mg F/kg soil)	63-d feeding/10-d F balance trial	43–44		Grace et al. (2005)
Sheep (2–5 yr old)	Seven soils (30% by wt. in forage feed, 235–1,030 mg F/kg soil, close to an Al smelter)	28-d feeding/5-d F balance trial	5–25		Milhaud et al. (1989)
Lambs (at weaning)	Dicalcic phosphate + grain meal	70-d feeding/bones analyzed		50–53[a]	Clay and Suttie (1985)
	Defluorinated phosphate			18–21[a]	
	Phosphate rock			61–72[a]	
Bulls (5 yr old)	Soil (6% by wt in hay + grain feed, 718 mg F/kg soil, close to an Al smelter)	10-mon feeding/5 balance studies	30–41		Wöhlbier et al. (1968)

[a]F retention relative to NaF.

Table 7. Effect of soil F feeding on the F concentrations (mg/kg dry matter) of ribs and radius bones, kidney, and liver of young sheep 63 d after treatment.

Treatment	Lucerne basal diet 7 mg F/kg DM	Lucerne + soil F 184 mg F/kg DM
Rib	739 ± 42.0	1,388 ± 45.9
Radius	726 ± 40.4	1,271 ± 76.5
Kidney	1.1 ± 0.4	2.9 ± 0.7
Liver	>0.08	>0.08

DM, dry matter.
Source: Grace et al. (2003).

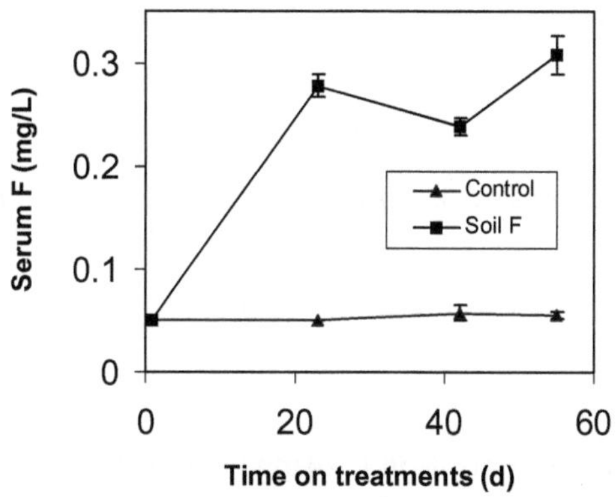

Fig. 14. Changes in blood serum F concentrations of young sheep with time after soil F feeding commenced. (Redrawn from Grace et al. 2003.)

in sheep fed with soil increased significantly within 23 d and then remained constant up to 55 d (Fig. 14). The F intake had little or no effect on the F concentrations of the kidney or liver, but bone F concentrations were significantly increased by the end of the trial at 63 days (see Table 7). The bone F concentrations, however, were below a threshold bone F concentration of 2400 mg F/kg DM, at which minor bone abnormalities may be observed. This fact suggests that young sheep exposed to an increase in F intake, as a result of soil ingestion, for one winter season under New Zealand conditions are unlikely to develop chronic fluorosis (Grace et al. 2003). Further studies (Grace et al. 2007) have shown that some of this bone F may be mobilized and that bone F concentrations decreased when F intakes are lower during summer and autumn (Fig. 15).

Results from a 345-d soil feeding trial (Grace et al. 2007) in which lambs were daily fed for 94 d with lucerne/grain pellets containing soil

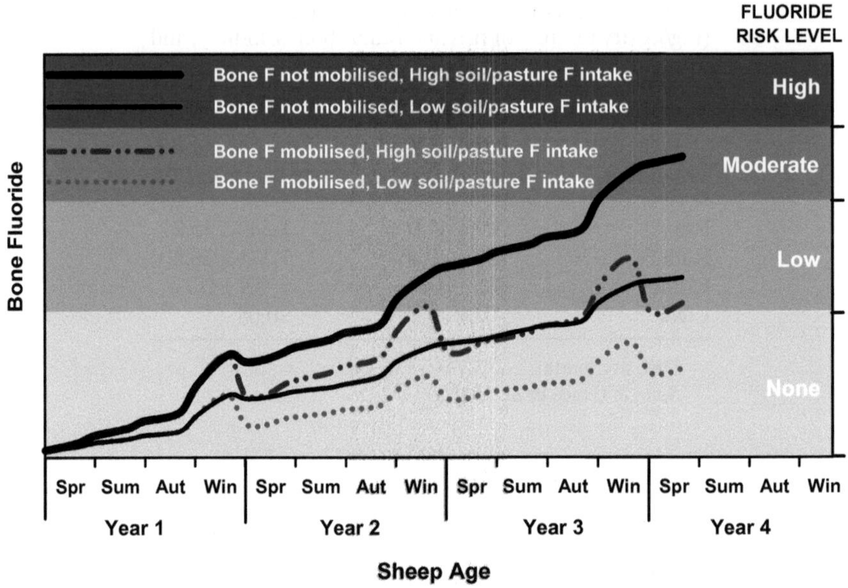

Fig. 15. Changes in fluorosis risk levels in grazing sheep for two F intake rates and F metabolism scenarios. Data for right-hand axis include the following: None, 200–2400; Low, 2400–3200; Moderate, 3200–5500; and High, 5500–7000 mg/kg DM. (From Shupe 1980; Wheeler and Fell 1983; Wheeler and Turner 1986.)

having very high F concentration (224 mg F/kg DM containing 10% soil by weight) showed that the bone F concentration increased from approximately 200 mg F/kg dry matter (DM) at day 1 to 1800–2700 mg/kg DM at day 94. When the F intake was decreased at day 94 (12 mg F/kg DM containing only lucerne/grain), the bone F concentrations then decreased to 1069–1075 mg/kg DM after 59 days (at day 153), and remained at not more than 1200–1500 mg/kg DM for up to day 345. These results show that a part of the F accumulated in the bones during winter/early spring months is remobilized when the F intake becomes low once again during summer and autumn. Further studies are required to investigate this cumulative seasonal effect of alternating high and low F intakes as the animal ages.

XI. Measures to Control Cadmium and Fluorine Accumulation in Soils, Pastures, and Livestock

A. Cadmium

An obvious method of reducing Cd accumulation in soils, pasture, and grazing animals is to reduce Cd input to the soil. In urban and industrialized areas, the atmospheric fallout of Cd is a major contributor to Cd input to soils. During the last decade this source of Cd has been significantly reduced by flue-gas purification. However, in pasture lands where P fertilizers are the major source, the Cd input can be achieved by reducing the Cd concentration in the P fertilizers applied to the soil or reducing P fertilizer application rates. Many countries in the world have progressively reduced the Cd concentration in fertilizers over the last decade (Table 8). This is achieved by blending PRs with different Cd concentrations.

Another method of reducing Cd concentration in fertilizers is by volatilizing most of the Cd by calcination of the PRs because Cd has a low boiling point (BP = 765°C). The manufacture of most phosphoric acid used in the food industry involves calcining PR before acidulation. However, calcination of PRs may not be a practical option in the fertilizer industry because it is expensive and calcination decreases the reactivity of PRs, making them unsuitable for direct application as a source of P. In some PRs, a significant amount of Cd is associated with the organic carbon present in the PRs (Loganathan and Hedley 1997). Therefore it may be possible to remove Cd through the removal of organic matter in the PRs using caustic solvents as these can remove the associated Cd without removal of P.

Table 8. Proposed or implemented Cd limits (legislative or voluntary) for P fertilizers in several countries (1994–2000).

Country	Cd limit, mg/kg P[a]	Effective year (in or before)
Australia	450	1994
	350	1995
	300	2000
Austria	170	2000
Belgium	207	2000
Czech Republic	116	2000
Denmark	150	1994
	110	2000
Finland	50	1994
Germany	200	1994
	93–207	2000
Japan	340	1994
Norway	100	1995
Sweden	100	1994
Switzerland	50	1994
New Zealand	420	1995–1996
	340	1997–1999
	280	2000

[a](mg Cd/kg P) × (% P in fertilizer) ÷ 100 = mg Cd/kg fertilizer.
Sources: Mortvedt (1996); Oosterhuis et al. (2000); Pallière (2005).

To increase pasture production, farmers often apply P fertilizers at rates higher than those required for optimum pasture production. Higher rates of P application not only reduces the efficiency of P utilization by plants but also can lead to P runoff to surface water bodies, resulting in adverse effects on the water quality. This problem can be minimized by only using P fertilizer rates that are required to reach an optimal soil test value. The P rate reductions on farms to reduce environmental problems will also reduce Cd input to the soils. For example, on a sheep farm at Whatawhata Research Centre, near Hamilton, New Zealand, a strong relationship was reported between total soil Cd concentration and soil test P (Olsen P) for the topsoils (Fig. 16). The Olsen P value recommended for optimum pasture production for this type of soil is 20–30 mg P/L (Morton and Roberts 1999). Figure 16 shows that in paddocks where this level of Olsen P is maintained soil Cd remained less than 0.15 mg Cd/kg soil. However, where excessive rates of P were applied resulting in levels of Olsen P higher than the optimum, the total soil Cd increased to values as high as 0.4 mg Cd/kg soil.

Alternative sources of P with low Cd/P ratio such as certain organic manures can be used to reduce Cd accumulation in soils. However, care should be taken in using composts made from municipal and urban wastes as they may have a high concentration of Cd (Chaney et al. 2006). In New Zealand, farm dairy effluent (FDE, excreta, and yard and milking shed washings) is commonly spray-irrigated to approximately 10%–20% of the

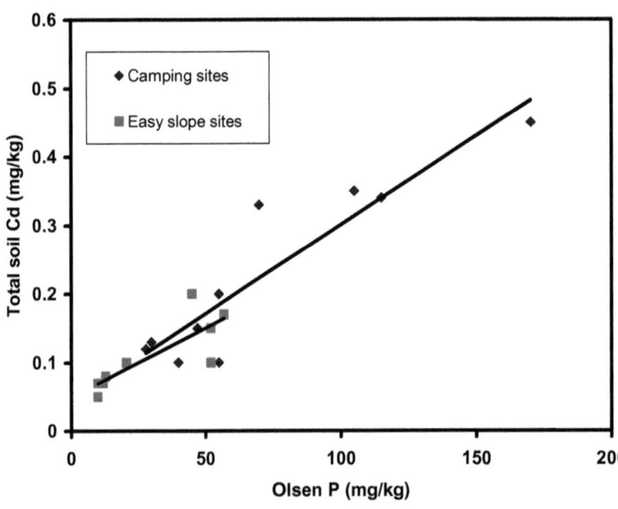

Fig. 16. Relationships between Olsen P (x) and total soil Cd concentration (y) in a sheep-grazed hill country pasture at Whatawhata research station (near Hamilton, New Zealand). Regression equations: animal camping sites, $y = 0.0026x + 0.0414$, $R^2 = 0.8412$; easy slope (<10%) $y = 0.002x + 0.0488$, $R^2 = 0.6882$. (Data sourced from Longhurst et al. 1994.)

farms grazing area as the preferred treatment of the FDE biological oxygen demand and nutrient load (Wang et al. 2004). As the effluent has a very low Cd/P ratio, in areas where effluent is applied the Cd input to soils is expected to be low.

Soil management practices that can reduce Cd accumulation in pasture and thereby reduce ingestion in animals include amelioration of soil acidity by lime application (increasing pH reduces plant availability of Cd), maintaining high levels of soil organic matter (to increase Cd sorption, thereby decrease plant availability of Cd), alleviating Zn deficiency (Zn reduces Cd plant uptake), and reducing weeds, which are known to have a much higher Cd concentration than grass and clover. However, contrasting results have been obtained on the effect of liming on plant availability of Cd in field conditions. Possible reasons for this have been discussed here.

Cultivation and resowing of permanent pasture reduces topsoil Cd concentration by diluting the higher Cd concentration in topsoil with subsoil containing lower Cd concentration. This practice decreases Cd intake by grazing animals through both soil ingestion and pasture uptake. During cultivation, lime can be incorporated into subsoil to reduce subsoil acidity, thereby reducing plant availability of subsoil Cd. Reducing stocking rates and avoidance of winter hard grazing can also reduce Cd uptake by soil ingestion.

Although measures need to be put in place to reduce the rate of increase in soil Cd, the metabolism of Cd in grazing animals is such that red meat and milk are very low in Cd and it is only the livers and kidneys that may increase the Cd intake of the population. However, as most meat-producing animals are slaughtered at a young age, this ensures that Cd concentrations of their edible offal products are within MPC criteria.

Supplementary Zn is sometimes administered to livestock for controlling facial eczema. Because Zn competes with Cd for binding to MT, suggestions have been made that Zn dosing may reduce Cd accumulation in animals. However, Zn treatments are typically administered over only a few weeks of the year, so the effectiveness of using this as a means to reduce the Cd burden of the whole animal over its lifetime may be limited. However, in young, rapidly growing animals, when rate of Cd accumulation is highest (Lee et al. 1996), high doses of Zn could be used to reduce liver Cd.

B. Fluorine

Currently there is no cost-effective method to reduce F concentration of P fertilizers, and therefore this is not an option for decreasing the rate of F accumulation in soils and its subsequent intake by animals. During the manufacture of P fertilizers by acidulation of phosphate rocks, the gaseous F compounds released are removed by scrubbing with water, and the resulting scrubber liquor is then often used in the fertilizer manufacturing process (acid dilution and fertilizer granulation). If the scrubber liquor is not recycled, the F concentration in the final fertilizer product can be reduced slightly.

As discussed in the preceding section on Cd, reducing P fertilizer application rates will reduce F soil input. Increases in soil test values (e.g., Olsen P) are strongly correlated with increased total soil F concentration (Loganathan et al. 2003), as was observed for Cd (see Fig. 16). Therefore, by not exceeding the optimum soil test value, farmers can avoid applying excess P, which will slow the rate of F input to soils. Furthermore, where available, organic manures with low F/P ratio can also be used to reduce the F input.

Withholding stock from pastures recently top-dressed with P fertilizers until at least 25 mm of rain has fallen to wash away the F can prevent acute F poisoning from fertilizer ingestion. This step requires a planned fertilizer application program so that some unfertilized paddocks are always available for grazing, or livestock can be fed forage crops until rain washes down the fertilizers adhering to the pasture. Under rotational grazing, fertilizer can be spread on paddocks that have been most recently grazed to maximize the interval between application and grazing.

As soil ingestion is the main source of F intake by grazing livestock, it can be minimized by maintaining good pasture cover and reducing the stocking rates of grazing stock, especially during winter. Grazing livestock ingest topsoils, 0–30 mm depth, that generally have a higher F concentration than the subsoil horizons. Therefore deep-ploughing lands high in topsoil F concentration during pasture renovation, as proposed for Cd, can reduce topsoil F and hence F intake by animals. A recent study by Loganathan et al. (2007) showed that cultivation of soil by ploughing to a depth of 150 mm using a mouldboard plough once in 6–9 yr decreased F in the top 30 mm by approximately 20%.

Summary

Fertilizers are indispensable for ensuring sustainability of agricultural production, thereby achieving food and fiber security. Nitrogen, sulfur, and potassium fertilizers are relatively free of impurities, but phosphorus (P) fertilizers, the main fertilizer input for the economic production of legume-based pastures, contain several contaminants, of which F and Cd are considered to be of most concern because they have potentially harmful effects on soil quality, livestock health, and food safety. Incidences of fluorosis in grazing livestock, and accumulation of Cd in the edible offal products of livestock, above the maximum permissible concentration set by food authorities have been reported in many countries. The majority of Cd and F applied to pastures in many countries continues to accumulate in the biologically active topsoil due to strong adsorption to soil constituents. However, the rate of Cd accumulation in the last decade has slowed as a result of selective use of low-Cd fertilizers.

Cd and F adsorption in soils increase with increased contents of iron and aluminium oxides, layer silicates and allophane in soils, and increased soil pH. Cadmium adsorption also increases with increased Mn oxides and organic matter in soil. However, some Cd will be released during decomposition of plant and animal remains and organic matter. In most pastoral soils the majority of Cd and F added in fertilizers remains in the topsoil and little moves below 20–30 cm, and therefore these are unlikely to contaminate groundwater. However, F may pose a risk to shallow groundwater in very acidic low-P-fixing soils, and Cd may pose a risk in very acidic soils containing low organic matter and clay contents, or in soils with high chloride concentrations. Research is required both to test whether groundwater beneath farms with long histories of P fertilizer use is contaminated by these elements and also to examine their mechanisms of movement.

Cd intake by grazing livestock occurs mostly by ingestion of pasture, and therefore measures to decrease plant availability of Cd in soils (e.g., maintaining high organic matter, reducing soil acidity and salinity, alleviating zinc deficiency, reducing weed) can reduce Cd accumulation in livestock. F intake by grazing livestock is mostly by soil ingestion; therefore, reducing soil ingestion by maintaining good pasture cover especially during winter periods can reduce F accumulation in livestock. In grazing livestock, Cd accumulates in kidney and liver, whereas F accumulates mainly in bones.

Very little research has been carried out to study the effects of sustained but low levels of Cd and F additions on soil microbial activity, especially on the economically important N-fixers in symbiosis with pasture legumes and mycorrhizae. This subject also needs to be researched.

The impact of F accumulation in bones of animals as influenced by the alternative low and high soil F intake between seasons and the effect of increasing age of animals needs further study to more accurately determine the potential risk of fluorosis and elucidate potential solutions to minimize F accumulation in bones and teeth of aged breeding stock.

Computer-based models are required to identify farming systems that present a high risk of Cd concentrations in edible offal exceeding the MPC and livestock at risk of chronic fluorosis. A decision support model of this kind may be useful in developing management strategies capable of reducing Cd and F accumulation in animals. Preliminary empirical models have been developed for Cd (Loganathan et al. 1999) and F (Bretherton et al. 2002) accumulation in sheep grazing New Zealand pastures. Further development of these models is required for their wider applicability.

Acknowledgments

We thank Mike Bretherton and Glenys Wallace of the Fertilizer and Lime Research Centre, Massey University, New Zealand for assistance with obtaining data on F and Cd accumulation in soil, plant and animal tissues, and Dr. Manuel Bermudez for assisting with drawing figures.

References

Adriano DC (2001) Trace Elements in the Terrestrial Environments: Biogeochemistry, Bioavailability, and Risks of Metals, 2nd Ed. Springer-Verlag, New York.

Arnesen AKM (1997) Availability of fluoride to plants grown in contaminated soils. Plant Soil 191:13–25.

Boekhold AE, Van der Zee (1992) SEATM: a scaled sorption model validated at the column scale to predict cadmium contents in a spatially variable field. Soil Sci 154:105–112.

Bolan NS, Naidu R, Syers JK, Tillman RW (1999) Surface charge and solute interactions in soils. Adv Agron 67:88–141.

Bolan NS, Adriano DC, Naidu R, Luz Mora Mdl, Mahimairaja S (2005) Phosphorus-trace element interactions in soil-plant systems. In: Sims JT, Sharpley AN (eds) Phosphorus: Agriculture and the Environment. Agronomy Monograph No. 46. American Society of Agronomy, Madison, WI, pp 317–352.

Bower CA, Hatcher JT (1967) Adsorption of fluoride by soils and minerals. Soil Sci 103:151–154.

Bremner I (1978) Cadmium toxicity. World Rev Nutr Diet 32:165–197.

Bretherton MR, Loganathan P, Hedley MJ, Grace ND (2002) A model to assess the risk of chronic fluorosis in sheep grazing New Zealand pastures. Report to New Zealand Fertiliser Manufacturers' Research Association, Auckland.

Brown PH, Dunemann L, Schulz R, Marschner H (1989) Influence of redox potential and plant species on the uptake of nickel and cadmium from soils. Z Pflanzenernaehr Bodenkd 152:85–91.

Cameron RS, Ritchie GSP, Robson AD (1986) Relative toxicities of inorganic aluminium complexes to barley. Soil Sci Soc Am J 50:1231–1236.

Castilho DP, Bril J, Romkens P, Oenema O (1996) Cadmium accumulation and availability in agricultural land and the effects of land use changes. In: Fertilizers as a Source of Cadmium. Organisation for Economic Cooperation and Development, Paris, pp 153–167.

Chaney RL, Filcheva E, Green CE, Brown SL (2006) Zn deficiency promotes Cd accumulation mby lettuce from biosolids amended soils with high Cd : Zn ratio. J Residuals Sci Technol 3:79–85.

Chien SH, Carmona G, Prochnow LI, Austin ER (2003) Cadmium availability from granulated and bulk-blended phosphate-potassium fertilizers. J Environ Qual 32:1911–1914.

Clark RG, Stewart DJ (1983) Fluorine (F). In: Grace ND (ed) The Mineral Requirements of Grazing Ruminants. New Zealand Society of Animal Production. THL Publishers, Singapore, pp 129–134.

Clark RG, Hunter AC, Stewart DJ (1976) Deaths in cattle suggestive of subacute fluorine poisoning following the ingestion of superphosphate. N Z Vet J 24:193–197.

Clay AB, Suttie JW (1985) The availability of fluoride from NaF and phosphorus supplements. Vet Hum Toxicol 27:3–6.

Cook PJ, Freney RJ (1988) Sources of cadmium in agriculture. In: Simpson J, Curnow WJ (eds) Cadmium Accumulations in Australian Agriculture. Bureau of Rural Resources, Canberra, ACT, pp 4–19.
Coote GE, Cutress TW, Suckling GW (1997) Uptake of fluoride into developing sheep teeth, following the 1995 volcanic eruption of Mt. Ruapehu, New Zealand. Nucl Instrum Methods Phys Res B 130:571–575.
Cronin SJ, Manoharan V, Hedley MJ, Loganathan P (2000) Fluoride: a review of its fate, bioavailability, and risks of fluorosis in grazed-pasture systems in New Zealand. N Z J Agric Res 43:295–321.
Cronin SJ, Neall VE, Lecointre JA, Hedley MJ, Loganathan P (2003) Environmental hazards of fluoride in volcanic ash: a case study from Ruapehu volcano, New Zealand. J Volcanol Geotherm Res 121:271–291.
Davister A (1996) Studies and research on processes for the elimination of cadmium from phosphoric acid. In: Fertilizers as a Source of Cadmium. Organisation for Economic Cooperation and Development, Paris, pp 21–30.
Degryse F, Smolders E (2006) Mobility of Cd and Zn in polluted and unpolluted Spodosols. Eur J Soil Sci 57:122–133.
FAO Statistics (2006) Landuse 2003. http://faostat.fao.org/site/418/default.aspx.
Fergusson JE, Stewart C (1992) The transport of air-bourne trace elements copper, lead, cadmium, zinc and manganese from a city into rural area. Sci Total Environ 121:247–269.
Fleischer M, Robinson WO (1963) Some problems of the geochemistry of fluorine. In: Shaw DM (ed) Studies in Analytical Geochemistry. Royal Society of Canada Spec Publ No. 6. University of Toronto Press, Toronto, pp 58–75.
Fung KF, Wong MH (2004) Application of different forms of calcium to tea soil to prevent aluminium and fluorine accumulation. J Sci Food Agric 84:1469–1477.
Furness H (2001) Cadmium in New Zealand. Its Presence and Management. New Zealand Fertiliser Manufacturers' Research Association, Auckland.
Grace ND, Loganathan P, Hedley MJ, Wallace GC (2003) Ingestion of soil fluorine: its impact on the fluorine metabolism and status of grazing young sheep. N Z J Agric Res 46:279–286.
Grace ND, Loganathan P, Deighton MW, Molano G, Hedley MJ (2005) Ingestion of soil fluorine: its impact on the fluorine metabolism of dairy cows. N Z J Agric Res 48:23–27.
Grace ND, Loganathan P, Hedley MJ (2007) The effect of temporal change in ingestion rates of fluorine (F) in soil on the concentration of F in serum and bone of young sheep. N Z Vet J 55:77–80.
Gray CW, McLaren RG, Roberts AHC, Condron LM (1998) Sorption and desorption of cadmium from some New Zealand soils: effect of pH and contact time. Aust J Soil Res 36:199–216.
Gray CW, McLaren RG, Roberts AHC, Condron LM (1999a) The effect of long-term phosphatic fertiliser applications on the amounts and forms of cadmium in soils under pasture in New Zealand. Nutr Cycl Agroecosyst 54:267–277.
Gray CW, McLaren RG, Roberts AHC, Condron LM (1999b) Solubility, sorption and desorption of native and added cadmium in relation to properties of soils in New Zealand. Eur J Soil Sci 50:127–137.
Gray CW, McLaren RG, Roberts AHC (2003) Cadmium leaching from some New Zealand pasture soils. Eur J Soil Sci 54:159–166.

Healy WB (1968) Ingestion of soil by dairy cows. N Z J Agric Res 11:487–499.

Healy WB (1973) Nutritional aspects of soil ingestion by grazing animals. In: Butler GW, Bailet RW (eds) Chemistry and Biochemistry of Herbage, vol. 1. Academic Press, London, pp 567–588.

Hewitt AE (1998) New Zealand Soil Classification. Manaaki Whenua Press, Landcare Research New Zealand Ltd, Lincoln, New Zealand.

Jackson AP, Alloway BJ (1992) The transfer of cadmium from agricultural soils to the human food chain. In: Adriano DC (ed) Biogeochemistry of Trace Elements. Science Reviews, Chicago, pp 109–158.

Jones JG, Jones JM (1962) Suspected poisoning of a dairy herd by basic slag. Ir Vet J 16: 43–44.

Kau PMH, Smith DW, Binning P (1998) Experimental sorption of fluoride by kaolinite and bentonite. Geoderma 84:89–108.

Koh TS, Bansemer PC, Frensham AB (1998) A survey of the cadmium concentration in kidney, liver and muscle of South Australian cattle. Aust J Exp Agric 38:535–540.

Langlands JP, Donald GE, Bowles JE (1988) Cadmium concentrations in liver, kidney and muscle in Australian sheep and cattle. Aust J Exp Agric 28:291–297.

Larsen S, Widdowson AE (1971) Soil fluorine. J Soil Sci 22:210–222.

Lee J, Grace ND, Martell S (1991) Effect of high and sustained zinc supplements on trace element metabolism in sheep. Proc N Z Soc Anim Prod 51:173–177.

Lee J, Treloar BP, Grace ND (1994) Metallothionein and trace element metabolism in sheep tissues in response to high and sustained zinc doses. II. Expression of metallothionein mRNA. Aust J Agric Sci 45:321–332.

Lee J, Rounce JR, Mackay AD, Grace ND (1996) Accumulation of cadmium with time in Romney sheep grazing ryegrass-white clover pasture: effect of cadmium from pasture and soil intake. Aust J Agric Sci 47:877–894.

Loganathan P, Hedley MJ (1997) Downward movement of cadmium and phosphorus from phosphatic fertilisers in a pasture soil in New Zealand. Environ Pollut 95:319–324.

Loganathan P, Mackay AD, Lee J, Hedley MJ (1995) Cadmium distribution in hill pastures as influenced by 20 years of phosphate fertilizer application and sheep grazing. Aust J Soil Res 33:859–871.

Loganathan P, Hedley MJ, Gregg PEH, Currie LD (1996) Effect of phosphate fertiliser type on the accumulation and plant availability of cadmium in grassland soils. Nutr Cycl Agroecosyst 46:169–179.

Loganathan P, Louie K, Lee J, Hedley MJ, Roberts AHC, Longhurst RD (1999) A model to predict kidney and liver cadmium concentration in grazing animals. N Z J Agric Res 42:423–432.

Loganathan P, Hedley MJ, Wallace GC, Roberts AHC (2001) Fluoride accumulation in pasture forages and soils following long-term applications of phosphorus fertilizers. Environ Pollut 115:275–282.

Loganathan P, Hedley MJ, Grace ND, Lee J, Cronin SJ, Bolan NS, Zanders JM (2003) Fertiliser contaminants in New Zealand grazed pasture with special reference to cadmium and fluorine: a review. Aust J Soil Res 41:501–532.

Loganathan P, Gray CW, Hedley MJ, Roberts AHC (2006) Total and soluble fluorine concentrations in relation to properties of soils in New Zealand. Eur J Soil Sci 57:411–421.

Loganathan P, Bretherton MR, Hedley MJ (2007) Effect of soil cultivation and winter pugging on fluorine distribution in soil profiles under pasture following long-term applications of phosphate fertilisers. Aust J Soil Res 45:41–47.

Longhurst RD, Roberts AHC, Brown MW, Carlson B (1994) Cadmium cycling in sheep-grazed hill country pastures. In: Currie LD, Loganathan P (eds) The Efficient Use of F in a Changing Environment: Reconciling Productivity with Sustainability. Fertilizer and Lime Research Centre, Massey University, Palmerston North, New Zealand, pp 297–302.

Louekari K (1996) The relation between the cadmium content in soil and in food plants. In: Fertilizers as a Source of Cadmium. Organisation for Economic Cooperation and Development, Paris, pp 180–188.

MacLean DC, Hansen KS, Schneider RE (1992) Amelioration of aluminium toxicity in wheat by fluoride. New Phytol 121:81–88.

Manoharan V, Loganathan P, Parfitt RL, Tillman RW (1996) Changes in soil solution composition and aluminium speciation under legume-based pastures in response to long-term phosphate fertiliser applications. Aust J Soil Res 34:985–998.

Manoharan V, Loganathan P, Tillman RW, Parfitt RL (2007) Interactive effects of soil acidity and fluorine on soil solution aluminium chemistry and barley (*Hordeum vulgare* L.) root growth. Environ Pollut 145:778–786.

Marshall B (1993) Recent MAF cadmium residue data. In: Lee J (ed) Second Cadmium Research Liaison Meeting Proceedings. AgResearch Grasslands Research Centre, Palmerston North, New Zealand, pp 2–6.

McBride M, Sauve S, Hendershot W (1997) Solubility control of Cu, Zn, Cd and Pb in contaminated soils. Eur J Soil Sci 48:337–346.

McLaughlin MJ (2000) Unwanted passengers in fertilizers, a threat to sustainability? In: Currie LD, Loganathan P (eds) Soil Research: A Knowledge Industry for Land-Based Exporters. Occasional report no. 13. Fertilizer and Lime Research Centre, Massey University, Palmerston North, New Zealand, pp 7–28.

McLaughlin MJ, Tiller KG, Naidu R, Stevens DP (1996a) Review: The behaviour and environmental impact of contaminants in fertilizers. Aust J Soil Res 34:1–54.

McLaughlin MJ, Tiller KG, Hamblin A (1996b) Managing cadmium contamination of agricultural land. In: Fertilizers as a Source of Cadmium. Organisation for Economic Cooperation and Development, Paris, pp 189–217.

McLaughlin MJ, Simpson PG, Fleming N, Stevens DP, Cozens G, Smart MK (1997a) Effect of fertilizer type on cadmium and fluorine concentrations in clover herbage. Aust J Exp Agric 37:1019–1026.

McLaughlin MJ, Tiller KG, Smart MK (1997b) Speciation of cadmium in soil solutions of saline/sodic soils and relationship with cadmium concentrations in potato tubers (*Solanum tuberrosum* L.) Aust J Soil Res 35:183–198.

McLaughlin MJ, Hamon RE, McLaren RG, Speir TW, Rogers SL (2000) Review: A bioavailability-based rationale for controlling metal and metalloid contamination of agricultural land in Australia and New Zealanad. Aust J Soil Res 38:1037–1086.

McLaughlin MJ, Stevens DP, Keerthisinghe DG, Cayley JWD, Ridley AM (2001) Contamination of soil with fluoride by long-term application of superphosphates to pastures and risk to grazing animals. Aust J Soil Res 39:627–640.

Merry RH (1988) Investigations on cadmium in South Australia: rainfall, soils, cereals, pastures and soil-plant relations. In: Simpson J, Curnow WJ (eds)

Cadmium Accumulations in Australian Agriculture. Bureau of Rural Resources, Canbera, ACT, pp 62–79.
Milhaud GE, Clauw M, Joseph-Enriquez B (1989) Bioavailability in soil fluoride in sheep. Fluoride 22:188–194.
Milhaud GE, Charles E, Loubière M, Kolf-Clauw M, Joubert C (1992) Effects of fluoride on secretory and postsecretory phases of enamel formation in sheep molars. Am J Vet Res 53:1241–1247.
Moore CS, Ritchie GSP (1988) Aluminium speciation and pH of an acid soil in the presence of fluoride. J Soil Sci 39:1–8.
Morton JD, Roberts AHC (1999) Fertilizer use on New Zealand sheep and beef farms. New Zealand Fertilizer Manufacturers' Research Association and AgResearch, Auckland, New Zealand.
Mortvedt JJ (1996) Heavy metal contaminants in inorganic and organic fertilisers. Fert Res 43:55–61.
Naidu R, Bolan NS, Kookana RS, Tiller KG (1994) Ionic strength and pH effects on the adsorption of cadmium and the surface charge of soils. Eur J Soil Sci 45:419–429.
Nicholson FA, Jones KC, Johnston AE (1996) Evidence for the leaching of surface deposited cadmium in agricultural soils. In: Fertilizers as a Source of Cadmium. Organisation for Economic Cooperation and Development, Paris, pp 218–231.
Öborn I, Jansson G (1998) Effects of liming on cadmium contents of spring wheat and potatoes. PP 1–7. Can be obtained in internet CD-ROM. In: Proceedings of the 16th World Congress of Soil Sciences on CD-ROM, August 20–26, Montpellier, France. Windows Internet Explorer.
O'Hara PJ, Cordes DO (1982) Superphosphate poisoning of sheep: a study of natural outbreaks. N Z Vet J 30:153–155.
O'Hara P, Fraser AJ, James MP (1982) Superphosphate poisoning of sheep: the role of fluoride. N Z Vet J 30:199–201.
Omueti JAI, Jones RL (1977) Fluoride adsorption by Illinois soils. J Soil Sci 28:564–572.
Oosterhuis FH, Brouwer FM, Wijnants HJ (2000) A possible EU wide charge on cadmium in phosphate fertilisers: economic and environmental implications. Final report to the European Commission. Report number E-00/02. April 2000.
Pallière C (2005) Prospective EU Cadmium Regulation for Fertilizers. Phosphate 2005 Conference, Paris, 3–5 April 2005.
Parker DR, Chaney RL, Norvell WA (1994) Chemical equilibrium models: applications to plant nutrition research. In: Loeppert RH, Schwab AP, Goldberg S (eds) Chemical Equilibrium and Reaction Models. Soil Science Society of America Spec Publ No. 42. American Society of Agronomy, Madison, WI, pp 163–200.
Petering DH, Fowler BA (1986) Discussion summary. Roles of metallothionein and related proteins in metabolism and toxicity, problems and perspectives. Environ Health Perspect 65:217–224.
Peterson MG, Mercer JFB (1988) Differential expression of four linked sheep metallothionein genes. Eur J Biochem 174:425–429.
Petterson DS, Masters HG, Speijers EJ, Williams DE, Edwards JR (1991) Accumulation of cadmium in the sheep. 26-13–26-14 In: Momčilović B (ed) Trace Elements in Man and Animals 7. IMI, Zagreb, Yugoslavia.
Phillips CJC, Chiy PC, Omed H, Saatci M (1997) Cadmium pollution of feed consumed by sheep and its amelioration with zinc supplements. Proc Br Soc Anim Sci 149.

Pickering WF (1985) The mobility of soluble fluoride in soils. Environ Pollut B 9:281–308.
Prankel SH, Nixon RM, Phillips CJC (2005) Implications for the human food chain of models of cadmium accumulation in sheep. Environ Res 97:348–358.
Rayment GE (1988) Cadmium in Queensland's primary industries, 1988. In: Simpson J, Curnow WJ (eds) Cadmium Accumulations in Australian Agriculture. Bureau of Rural Resources, Canberra, ACT, pp 151–160.
Roberts AHC, Longhurst RD (2002) Cadmium cycling in sheep-grazed hill-country pastures. N Z J Agric Res 45:103–112.
Roberts TL, Stauffer MD (1996) Agronomic implications of restricting cadmium content of phosphate rock. In: Fertilizers as a Source of Cadmium. Organisation for Economic Cooperation and Development, Paris, pp 70–78.
Roberts AHC, Longhurst RD, Brown MW (1994) Cadmium status of soils, plants, and grazing animals in new Zealand. N Z J Agric Res 37:119–129.
Römkens PFAM, Vries WD (1995) Acidification and metal mobilization: effects of land use changes on cadmium mobility. In: Hey GJ, Erisman JW (eds) Acid Rain Research: Do We Have Enough Answers. Elsevier S, Amsterdam, pp 367–380.
Rounce JR, Knowles SO, Grace ND, Lee J (1998) Effect of zinc oxide treatment for facial eczema on the copper status of Romney sheep grazing ryegrass pastures. Proc N Z Soc Anim Prod 58:199–210.
Sauerbeck D (1992) Conditions controlling the bioavailability of trace elements and heavy metals derived from phosphate fertilizers in soils. In: Proceedings of the 4th International Imphos Conference on Phosphorus, Life and Environment. Institute Mondial du Phosphate, Casablanca, pp 419–448.
Shupe DVM (1980) Clinicopathologic features of fluoride toxicosis in cattle. J Anim Sci 51:746–758.
Simons JH (1950) Fluorine Chemistry. Academic Press, New York.
Soler JS, Rovira JS (1996) Cadmium in inorganic fertilizers. In: Rodriguez-Barrueco C (ed) Fertilizers and Environment. Kluwer, Dordrecht, pp 541–545.
Stevens DP, McLaughlin MJ, Alston AM (1997) Phytotoxicity of aluminium–fluoride complexes and their uptake from solution culture by *Avena sativa* and *Lycopersicon esculentum*. Plant Soil 192:81–93.
Stevens DP, McLaughlin MJ, Randall PJ, Keerthisinghe G (2000) Effect of fluoride supply on fluoride concentrations in five pasture species: levels required to reach phytotoxic or potentially zootoxic concentrations in plant tissue. Plant Soil 227:223–233.
Taylor MD, Percival HJ (2001) Cadmium in soils and soil solutions from a transect away from a fertilser bin. Environ Pollut 113:35–40.
Wang HL, Magesan GN, Bolan NS (2004) An overview of the environmental effects of land application of farm effluents. N Z J Agric Res 47:389–403.
Weinstein LH, Davison A (2004) Fluorides in the Enivironment. CABI Publishing, CAB International, Oxon, UK.
Wheeler SM, Fell LR (1983) Fluorides in cattle nutrition. Nutr Abstr Rev 53:741–767.
Wheeler SM, Turner AD (1986) The bone fluoride of ewes and lambs in NSW. Proc Aust Soc Anim Prod 16:391–394.
Williams CH, David DJ (1973) The effect of superphosphate on the cadmium content of soils and plants. Aust J Soil Res 11:43–56.

Wöhlbier W, Oelschlager W, Gronbach G, Giessler H (1968) Die Resorption von Fluor durch Ochsen aus Erde und Flugstaub einer Aluminiumhütte Forschungsberung 14 der DFG Fluor-Wirkung. Fr Steiner, Wiesbaden.
Zanders JM (1998) Studies on the origin, distribution and mobility of cadmium in pastoral soils. PhD thesis. Massey University, New Zealand.
Zasoski RJ, Burau RG (1988) Sorption and sorptive interaction of cadmium and zinc on hydrous manganese oxide. Soil Sci Soc Am J 52:81–87.

Manuscript received November 6; accepted November 8, 2006.

Rev Environ Contam Toxicol 192:67–115

Aquatic Plants Exposed to Pharmaceuticals: Effects and Risks

Richard A. Brain, Mark L. Hanson, Keith R. Solomon, and Bryan W. Brooks

Contents

Communicated by David H. Whitacre.

R.A. Brain (✉), B.W. Brooks
Center for Reservoir and Aquatic Systems Research, Department of Environmental Studies, Baylor University, One Bear Place Waco, Tx, USA, 76798–7388.

M.L. Hanson
Department of Environment & Geography, University of Manitoba, Dafoe Rd. Winnipeg, Manitoba, Canada, R3T 2N2.

K.R. Solomon
Department of Environmental Biology, University of Guelph, 50 Stone Rd. E. Guelph, Ontario, Canada N1G 2W1.

I. Introduction

The discovery of therapeutic medicinal products has historically been largely accidental, through trial and error involving observation of the therapeutic effects of "naturally" produced compounds such as penicillin from the mold *Penicillium notatum* (Fleming 1929), salicylic acid from the bark of willow trees (*Salix*) (Mackowiak 2000), and lovastatin from the fungal microorganism *Aspergillus terreus* (Bach and Lichtenthaler 1982a), with little or no preconceived knowledge of effect. Currently, however, the pharmacokinetics and pharmacodynamics of these compounds are well understood. However, amid all the research and development of synthetic analogues and derivatives, the original functions of these compounds, as they relate to the organisms in which they were discovered, appear to have been forgotten. The consequence, however, becomes readily apparent when considering the ecotoxicology of pharmaceuticals, particularly because these compounds are biologically active and produced to affect other organisms. Because there are numerous examples of evolutionary conservation of pathways and receptor targets within and among levels of biological organization, it stands to reason that pharmaceuticals derived, synthesized, and inspired from these naturally produced compounds could ultimately induce toxic effects in nontarget organisms. Recent research supports this notion, and this review discusses potential evolutionary conserved targets to, general effects from, and the risks posed by pharmaceuticals in nontarget aquatic plants.

II. Exposure of Aquatic Plants to Pharmaceuticals in the Environment

Interest related to pharmaceuticals in the environment is not new, as reports concerning antimicrobials and veterinary medicines date as far back as 1969 (Swann et al. 1969). Pharmaceuticals currently represent a diverse group of

>4,500 biologically active parent compounds (Sweetman 2002), with annual usage on a par with agrochemicals (Daughton and Ternes 1999) entering the environment via two main routes, sewage effluent from human sources and agricultural runoff from concentrated animal feeding operations and disposal of manure (Daughton and Ternes 1999; Halling-Sørensen et al. 1998). Human drugs tend to enter the environment from point sources, whereas the discharge of veterinary drugs tends to be more diffuse in nature.

The most significant entry route for human pharmaceuticals into the aquatic environment is the release from wastewater treatment plants (WWTPs). Despite apparent removal of a large proportion of many pharmaceuticals through various sewage treatment processes (Boyd et al. 2003), WWTPs do not eliminate all compounds completely and, as a consequence, pharmaceuticals are discharged into receiving waters in detectable amounts (Heberer 2002).

Animal husbandry operations, including veterinary applications, prophylaxis, and growth promotion, discharge drugs and their metabolites into the environment through liquid manure and (waste) water or with storm water runoff from fields after manure application (Kümmerer 2001). Veterinary medicines provide the greatest array of possibilities for exposure and can be subdivided into substances used as growth promoters for swine, as therapeutics in livestock production, cocciodiostatics used for poultry production, therapeutics for treatment of livestock on fields (e.g., antiparasitic agents), or into feed additives in aquaculture (Halling-Sørensen and Jørgensen 2000). Antibiotics used in aquaculture are directly added to receiving waters, formulated as feed additives (Halling-Sørensen et al. 1998), with 70%–80% of administered drug entering the environment (Samuelsen et al. 1992). In Europe, approximately 10,000 t of antibiotics are consumed per year, 50% for both veterinary and human purposes (FEDESA 1997), which is similar to the United States, where roughly 40% of the 22,680 t antibiotics produced each year is used in agriculture (Levy 1998). The continual release of pharmaceuticals allows many to take on a pseudo-persistence, displaying similar exposure characteristics as many truly persistent compounds, because the rate of replacement can balance the rate of transformation or removal in the environment (Daughton 2002). Such an observation is most commonly encountered in effluent-dominated systems (Brooks et al. 2006).

The first detection of pharmaceutical residues in the aquatic environment was reported as early as the 1970s (Garrison et al. 1976; Tabak and Brunch 1970). Recently, a national reconnaissance of U.S. streams, screening for 95 pharmaceuticals, hormones, and organic wastewater contaminants, frequently reported detection of several classes of pharmaceuticals. Measured concentrations were generally <1 μg/L, but concentrations of 5% of the 30 most frequently detected compounds exceeded 1 μg/L (Kolpin et al. 2002). A number of surveys have also been conducted in Canada identifying a variety of pharmaceutical compounds in the ng/L–μg/L range in surface waters receiving urban and agricultural inputs (Hao et al. 2006; Lissemore et al. 2006; Metcalfe et al. 2003; Miao and Metcalfe 2003; Miao et al. 2004; Sosiak and Hebben 2005). Similar surface water concentrations have been documented for a number of pharmaceuticals in Europe (Daughton and Ternes 1999; Halling-Sørensen et al. 1998; Heberer 2002; Hirsch et al. 1999; Jones et al. 2001; Kümmerer 2001) and, although there are diverse differences in demographic usage trends for pharmaceutical classes between countries, the overall residue concentration ranges are similar, suggesting that pharmaceuticals are present ubiquitously in the aquatic environment.

It was not until the late 1990s that the first comprehensive reviews regarding the effects of pharmaceuticals to nontarget aquatic organisms were published (Daughton and Ternes 1999; Halling-Sørensen et al. 1998). One major conclusion indicated by Daughton and Ternes (1999) was that the potential effects of pharmaceuticals and personal care products on nontarget species, especially on aquatic organisms, were mostly unknown, a conclusion echoed by Halling-Sørensen et al. (1998). Although few effect data for plants were available when these reviews were published, the potential for effects of pharmaceuticals, particularly antibiotics, in plants is certainly not a new concept. Reviews documenting the effects of antibiotics on plants have been compiled as far back as the 1950s (Brian 1957). However, these early studies focused largely on the applications of antibiotics for disease control in plants, although a number of antibiotic compounds were identified as phytotoxic. Investigations of the ecotoxicological effects of pharmaceuticals on aquatic plants are a much more recent area of research.

III. Aquatic Plants in the Environment and Their Use in Ecotoxicology

Photosynthetic organisms, both phytoplankton and macrophytes, represent a significant portion of the total biomass in aquatic environments and are important as a carbon source for the rest of the aquatic biosphere (Greenberg et al. 1992). Accounting for less than 1% of the earth's primary producer biomass, primary production from aquatic plants is estimated to be only slightly lower than terrestrial plants, 35–50 giga-tonnes versus 50–70, respectively (Falkowski and Raven 1997). Aquatic plants are also important components of aquatic ecosystems for a number of reasons (Davy et al. 2001): they contribute to primary productivity, generate oxygen, affect flow patterns of water, provide habitat and food for other organisms, stabilize sediment, are utilized by detritivors, are involved in nutrient cycling, and improve water quality. Thus, they are an appropriate assessment endpoint for ecotoxicology and environmental protection.

Comprehensive environmental risk assessment guidelines of pharmaceuticals are not yet fully established. Therefore, for the purposes of this review, the use of aquatic plants in the context of pesticide registration is provided as background. Current guidelines for pesticide registration provide test protocols for the tiered testing of one aquatic higher plant, *Lemna gibba*; a green algae, *Pseudokirchneria subcapitata*; a blue-green cyanobacteria, *Anabaena flosaquae*; a freshwater diatom, *Navicula pellicu-losa*; and a marine diatom, *Skeletonema costatum* (Davy et al. 2001). Under the present tiered approach for nontarget aquatic plant phytotoxicity, a limited number of species are tested in the first tier, and if the first tier tests show effects, then additional plant species are studied at higher tiers (U.S. EPA 2006). In the first tier, laboratory tests are conducted using *Pseu-dokirchneria subcapitata* and *L. gibba* to evaluate the acute toxicity of pes-ticides at the greatest application rate (U.S. EPA 2006). Second-tier testing involves concentration–response bioassays that are designed to evaluate the acute toxicity of pesticides to the five aquatic species listed above (U.S. EPA 2006). Finally, third-tier field studies may be required on a case-by-case basis if terrestrial plants show greater than 25% adverse effects on plant growth and aquatic plants show greater than 50% adverse effects on plant growth (U.S. EPA 2006). These tests provide critical information on harmful effects to plants during stages of development. However, aquatic and ter-restrial test species commonly used in plant toxicity tests cannot reliably serve as surrogates for all untested plants (Davy et al. 2001). As a result, revised guidelines for phytotoxicity have been developed and are being reviewed (Davy et al. 2001).

IV. Receptor Sites of Toxicity for Pharmaceuticals in Plants

Pharmaceuticals are biologically active compounds, designed to elicit a specific effect in humans or animals. Therefore, it may be expected that any effect, beneficial or adverse, could also occur in aquatic organisms with similar biological functions and receptors. A common question proposed in ecological risk assessment is whether appropriate receptors to a stressor exist in nontarget organisms, and in this context, it is necessary to consider evolutionary conservation of receptor systems. Many pharmaceuticals are expected to act via narcosis (Cleuvers 2003; Sanderson et al. 2004), although some compounds may elicit receptor-mediated responses if the organism contains the appropriate receptor (Seiler 2002). To obtain a respective response from a nontarget organism, the organism must possess the struc-ture or function that is targeted in the therapeutic indication, i.e., it must express the respective receptor or use the respective biosynthetic pathway, with enzymes exhibiting structures nearly identical, or at least as similar as possible, to their human counterparts (Seiler 2002).

When considering the question of whether some substances might harm the environment, and whether such influences could be predicted by the toxicological or pharmacological properties of a substance, a fundamental difference between ecotoxicological assessments and the conventional "anthropocentric" evaluation of the pharmacotoxicological profile of a drug has to be taken into account (Seiler 2002). Extrapolating effects that have been determined in the conventional pharmacodynamic and toxicological assays at large doses to the environmental situation with very small exposures may be inappropriate for most of the older and less-specific pharmaceuticals. Furthermore, the development of pharmaceuticals targeting more basic mechanisms of cellular functions may result in additional pathways for affecting nontarget organisms (Seiler 2002). Seiler (2002) further suggests that, at least for those drug substances that can influence their targets at in vitro EC_{50} values of around 1–10 nM, environmental concentrations in the range of 300 ng/L–5 μg/L should constitute effective target concentrations. This statement, however, is for aquatic exposure, and not the actual concentration reaching the target site in the organism (e.g., internal dose). The following sections outline potential or known receptors in aquatic plants to antibiotics and other classes of pharmaceuticals that are frequently prescribed and detected in the environment.

V. Evolutionary Conservation of Receptors and Pathways

Plants provide a number of evolutionary conserved target sites as a result of the bacterial ancestry of plastid organelles and conservation of certain metabolic pathways (Bach and Lichtenthaler 1983; Lichtenthaler et al. 1997a,b; McFadden and Roos 1999). All plastid organelles arose from the engulfment of an endosymbiotic cyanobacterial-like prokaryotic cell by a larger eukaryotic cell, referred to as the primary endosymbiotic event, generating a plastid characterized by two membranes, such as those found in red and green algae, plants, and glaucophytes (McFadden and Roos 1999). These plastids are contained within at least two bounding membranes, have DNA genomes that, like those of bacteria, are molecules of supercoiled circular double-stranded DNA, typically 120–160 kb in size, with approximately 130 genes in higher plants (Bock 2001), and a transcription and translation system (McFadden and Roos 1999). Although plastids have undergone substantial modification during an estimated 500 million years of intracellular survival within their eukaryotic hosts, the genome, transcription, and translation equipment of chloroplasts have remained fundamentally bacterial in nature (McFadden and Roos 1999). The cyanobacterial nature of plastid pathways and homology in replicating equipment therefore makes plastids susceptible as potential drug targets (McFadden and Roos 1999).

Recently, the entire *Arabidopsis thaliana* genome was sequenced and functionally compared to the complete genomes of *Escherichia coli*, *Synechocystis* sp., *Saccharomyces cerevisiae*, *Caenorhabditis elegans*, and *Drosophila*, and a nonredundant protein set of *Homo sapiens* using a basic local alignment search tool to query protein amino acid sequences (BLASTP) with a stringent expected threshold of $E < 10^{-30}$ (The Arabidopsis Genome

Initiative 2000). The proportion of *Arabidopsis* proteins having related counterparts in eukaryotic genomes was found to vary by a factor of 2–3 depending on the functional category, with 8%–23% relatedness of proteins involved in transcription and 48%–60% relatedness of genes involved in protein synthesis (The Arabidopsis Genome Initiative 2000). Furthermore, a relatively high proportion of matches was found between *Arabidopsis* and bacterial proteins in the categories "metabolism" and "energy," which is considered to reflect both the acquisition of bacterial genes from the plastid ancestor and high conservation of sequences across all species (The Arabidopsis Genome Initiative 2000). This degree of similarity, particularly among protein synthesis genes, indicates a high potential for functional conservation of enzymes and metabolic pathways across highly diverse groups of organisms and a corresponding potential for susceptibility among these enzymes and metabolic pathways as pharmacological targets.

A. Chloroplast Replication and DNA Gyrases in Plants as Targets for Fluoroquinolone Antibiotics

In bacteria, fluoroquinolone antibiotics act by inhibiting the activity of DNA gyrases, which is the bacterial equivalent of DNA topoisomerase (topo) that supercoils DNA by using the free energy of ATP hydrolysis (Champoux 2001; Divo et al. 1988). Topoisomerases are enzymes present in all cells that control the topological state of DNA, and there are two types, types I and II, distinguished by the ability to transiently break one or both strands of the DNA (Champoux 2001). DNA gyrase is a type II enzyme required for replication and transcription (Fig. 1) in prokaryotes, and is the only enzyme of this type that is able to catalyze ATP-dependent DNA supercoiling (Champoux 2001; Wall et al. 2004). Because gyrases are apparently absent from eukaryotes and are critical for prokaryotes, a number of antibacterial agents, including quinolones, coumarins, and cyclothialidines, have been developed to specifically target this enzyme (Maxwell 1997). Although DNA gyrases have previously been thought to be unique to bacteria, recent evidence indicates that plants contain DNA gyrases (Thompson and Mosig 1985; Wall et al. 2004).

In the mid-1980s, Thompson and Mosig (1985) demonstrated the weak activity of an ATP-dependent topoisomerase in the green alga *Chlamydomonas reinhardtii* that could supercoil DNA in vitro, and, although the enzyme could not be purified, they found that the gyrase-specific drug novobiocin (a coumarin) inhibited chloroplast transcription. More recently, Wall et al. (2004) identified four putative gyrase genes in the model plant *A. thaliana*, one gyrA and three gyrB homologues, confirming the existence of DNA gyrase in eukaryotes. The encoded DNA gyrase protein A (GyrA) was found to have a dual translational initiation site targeting the mature protein to both chloroplasts and mitochondria, whereas DNA gyrase protein B (GyrB) homologues were found to have individual chloroplast

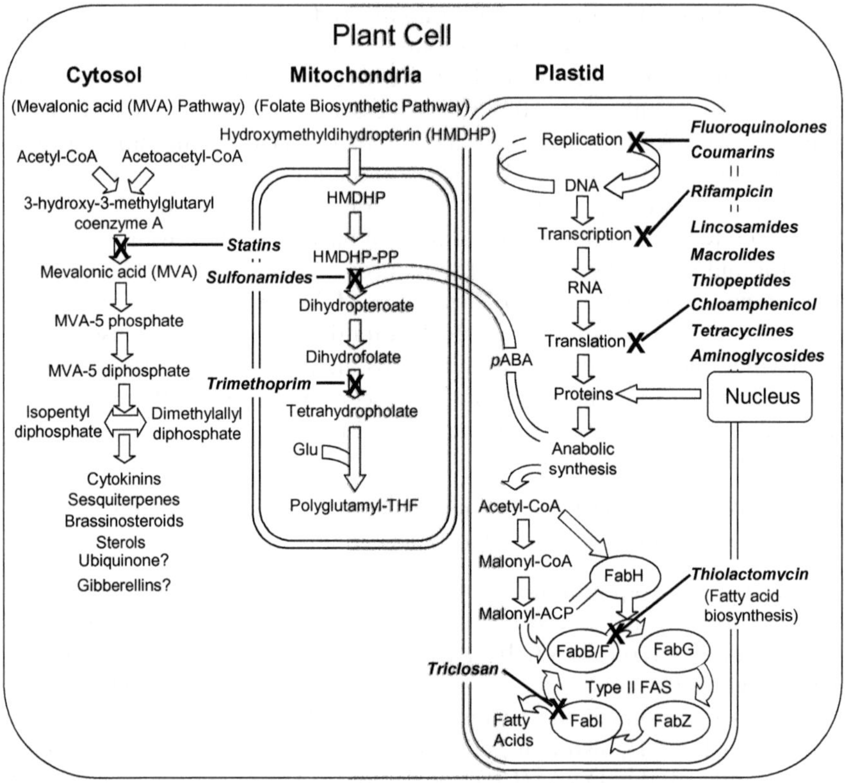

Fig. 1. Select metabolic pathways within the cytosol, mitochondria and plastid of a plant cell and the major classes of pharmaceuticals believed to target these pathways or processes. Statins are known inhibitors of 3-hydroxy-3-methylglutaryl coenzyme A reductase (HMGR) in the mevalonic acid pathway, sulfonamide antibiotics are thought to target dihydropteroate synthase (DHPS) in the folate biosynthetic pathway, a variety of antibiotic classes are suspected to inhibit specific processes within the chloroplast, such as replication, transcription, and translation, and triclosan and thiolactomycin are thought to disrupt fatty acid biosynthesis by inhibiting enoyl-ACP reductase (FabI) and β-ketoacyl-ACP synthase I and II (FabB and/or FabF), respectively. The context of each of these receptors/pathways is described in the text along with the definition of acronyms. Modified from (Basset et al. 2005; McFadden and Roos 1999; Schwender et al. 1996; Waller et al. 2003).

and mitochondria targeting sequences (Wall et al. 2004). Organelle targeting was confirmed by N-terminal fusions of the sequences to green fluorescence protein (GFP), demonstrating that one enzyme is targeted to the chloroplast and another to the mitochondrion, which correlated with supercoiling activity in isolated organelles (Wall et al. 2004). Furthermore, treatment of seedlings and cultured cells with ciprofloxacin caused growth inhibition, and knockouts of *A. thaliana* gyrA were found to be embryo lethal, whereas knockouts in the gyrB genes lead to seedling-lethal phenotypes or severely stunted growth and development (Wall et al. 2004).

Consistent with these findings is the robust dataset demonstrating antichloroplastic activity of quinolones and coumarins to the apicomplexan plastids (plastids in parasites of the phylum Apicomplexa) of a number of protists (Adams et al. 1974; Krajcovic et al. 1989; Lyman et al. 1967; McFadden and Roos 1999; Woods et al. 1996).

B. Translation and Transcription Processes in the Chloroplast as Targets for Tetracycline, Macrolide, Lincosamide, Aminoglycoside, and Pleuromutilin Antibiotics

The translational apparatus and mRNAs of plastids are similar to those in bacteria. Where chloroplasts are responsible for the transcription and translation of the genes encoded in their genome, the machinery used to accomplish this shares a high degree of homology with that found in bacteria (Manuell et al. 2004). Molecular phylogenetic analysis of c-type cytochrome and rRNA sequences have established a relationship between cyanobacteria and the green (euglenoids, green algae, and higher plants) and red (rhodophyte) chloroplasts, supporting the prokaryotic origins of chloroplasts (Giovannoni et al. 1988). Analysis of partial sequences of 16S rRNAs from 29 cyanobacteria and the cyanelle of the phytoflagellate *Cyanophora paradoxa* indicated that the cyanobacteria and green chloroplasts form a coherent phylogenetic group and that the chloroplast lineage is not just a sister group to the free-living forms but rather is contained within the cyanobacterial radiation (Giovannoni et al. 1988). Plastid genes contain consensus prokaryotic promoters and bacterial-like sigma factors and polymerases, and ribosome-binding sequences are found in many chloroplast mRNAs as a requirement for translation, presumably through a similar binding between mRNA and the ribosome as found in bacterial translation initiation (Manuell et al. 2004). A high degree of homology between the chloroplast and bacteria with regard to general translation factors and a majority of the ribosomal proteins implies that many of the basic processes of translation are conserved between bacteria and the chloroplast.

Plastids and mitochondria both have 70S ribosomes composed of two subunits, a 50S and a 30S, similar to prokaryotes. Yamaguchi et al. (2003) conducted a proteomic analysis of the 70S ribosome from the chloroplast of the green alga *C. reinhardtii* and identified 27 orthologues of *E. coli* large subunit proteins in the 50S subunit, as well as an orthologue of the spinach plastid-specific ribosomal protein (PSRP)-6, indicating that the large subunit proteins are very similar to those of spinach chloroplast and *E. coli*. LC-MS/MS analysis was used to identify the protein components of the *C. reinhardtii* chloroplast ribosome where 22 proteins from the small subunit were identified, including homologues of 20 of the 21 *E. coli* proteins (Yamaguchi et al. 2003). From the large subunit of the *C. reinhardtii* chloroplast ribosome, 28 proteins were identified, including homologues of 27 of the 33 *E. coli* proteins (Yamaguchi et al. 2003). One additional PSRP was identified having homology to PSRP-6, identified previously from the spinach chloroplast ribosome (Yamaguchi and Subramanian 2000).

Streptomycin, spectinomycin, neamine/kanamycin, and erythromycin resistance has been documented in the chloroplasts of *C. reinhardtii* and is conferred by base substitutions in conserved regions of the genes encoding the 16S and 23S chloroplast ribosomal RNAs (Harris et al. 1989). These mutants define eight genetic loci in a linear linkage group corresponding to about 21 kb of the circular chloroplast genome (Harris et al. 1989). Streptomycin resistance can result from base changes equivalent to *E. coli* 13, 523, and 912–915 in the 16S gene, which are involved in a common ribosomal neighborhood that interacts with ribosomal proteins S4, S5, and S12 (Harris et al. 1989). A point mutation at base 876 in *E. gracilis*, equivalent to base 912 in the 16S gene of *E. coli*, was also found to confer resistance to streptomycin (Montandon et al. 1985). Harris et al. (1989) also noted that three different changes within a conserved region of the 16S gene, equivalent to *E. coli* bases 1191–1193, confer varying levels of spectinomycin resistance, whereas resistance to neamine and kanamycin results from mutations in the 16S gene at bases equivalent to *E. coli* 1408 and 1409. Five mutations in two genetically distinct erythromycin resistance loci map in the 23S rDNA of *C. reinhardtii* were also identified at positions equivalent to *E. coli* 2057–2058 and 2611, and all were highly resistant to erythromycin, although they differed in levels of cross-resistance to lincomycin and clindamycin (Harris et al. 1989). As macrolide and lincosamide antibiotics block protein synthesis by interacting with the peptidyl tranferase domain of bacterial 23S rRNA, and aminoglycosides interfere with bacterial protein synthesis by irreversibly binding to 30S and 50S subunits of ribosomes, these compounds have the potential to disrupt transcription/translation in the chloroplasts of photosynthetic organisms (see Fig. 1).

Protein identification of the chloroplast ribosome (chloro ribosome) of higher plants has shown that the chloroplast ribosomal 30S subunit contains 4 chloroplast/PSRPs and the orthologues of the full complement of *E. coli* 30S subunit ribosomal proteins (designated PSRP 1–4) (Yamaguchi and Subramanian 2000, 2003). The chloroplast ribosomal 50S subunit is composed of 31 orthologues of the *E. coli* 50S ribosomal subunit proteins (Yamaguchi and Subramanian 2000, 2003). Investigations of intact chloroplast ribosome (70S) has revealed an additional protein, a plastid ribosome recycling factor (pRRF), which is released on the dissociation of the chloroplast ribosome into subunits (Yamaguchi and Subramanian 2000). Overall, the chloroplast ribosome proteome of higher plants is composed of 59 distinct proteins: 6 PSRPs, a bacterial-type pRRF, and 52 orthologues of eubacterial ribosomal proteins (Yamaguchi and Subramanian 2000). Six ribosomal proteins are specific to higher plant chloroplast ribosomes (Yamaguchi and Subramanian 2003).

At least two distinct DNA-dependent RNA polymerases have recently been identified through analyses of the plastidial transcription apparatus:

one of these is a multisubunit enzyme similar to those in bacteria and in the nuclei of eukaryotic cells, and the other enzyme is a single-subunit polymerase related to those of bacteriophages T3 and T7 and mitochondria (Gray and Lang 1998; Maliga 1998; Pfannschmidt et al. 2000). The multisubunit enzyme is referred to as plastid-encoded polymerase (PEP) because of the intraorganellar coding sites of its core subunits, whereas the phage-type enzyme is nuclear encoded (Gray and Lang 1998; Maliga 1998; Pfannschmidt et al. 2000). Plastid transcription utilizes an RNA polymerase (α_2, β, β') that is homologous to that of cyanobacteria and other eubacteria (Gray and Lang 1998). The plastid-encoded polymerase recognizes promoters with consensus –10 and –35 sequences via the σ-factor and transcribes polycistronic RNAs from plastid DNA operons (Mullet 1993). Two multisubunit plastid RNA polymerases termed A and B have been identified in mustard, where the B enzyme has a bacterial-type polypeptide composition and is sensitive to the prokaryotic transcription inhibitor rifampicin, and the A enzyme has a more complex subunit structure and is not sensitive (Pfannschmidt et al. 2000). However, the A enzyme can be converted into a rifampacin-sensitive enzyme form (Pfannschmidt et al. 2000). The bacterial nature of the plastidial RNA polymerase provides another potential target for antibiotics interfering with transcription/translation in photosynthetic organisms.

C. Mitochondrial Folate Biosynthesis in Plants as a Target for Sulfonamides

The mode of action for sulfonamide antibiotics is via the inhibition of the enzyme dihydropteroate synthase in the folate biosynthetic pathway acting as a structural analogue of the substrate *p*-aminobenzoic acid (*p*ABA), as shown in Fig. 1 (Sweetman 2002). Similarly, diaminopyrimidine antibiotics such as trimethoprim inhibit dihydrofolate reductase, two steps down from dihydropteroate synthase in the folate biosynthetic pathway also shown in Fig. 1. Folates are essential cofactors for one-carbon transfer reactions in most living organisms and are required for the biosynthesis of nucleic acids, amino acids, and pantothenate, and, specifically in plants, lignin formation and photorespiration (Basset et al. 2005; Hanson and Roje 2001). In contrast to bacteria and plants, humans cannot synthesize folates de novo, requiring assimilation from their diet (Basset et al. 2005). However, it has been reported that the plant folate synthesis pathway is essentially the same as in bacteria, and it has now been almost completely elucidated (Basset et al. 2005). Eight of the 10 enzymes specific to the folate biosynthetic pathway have been cloned and characterized from plants, and the metabolic steps of folate biosynthesis are outlined in Basset et al. (2005). Of consideration here is the enzyme dihydropteroate synthase, catalyzing the formation of dihydropteroate (precursor of tetrahydrofolate; see Fig. 1) from hydroxymethyldihydropterin pyrophosphate (HMDHP-PPi; produced in the cytosol) and *p*-aminobenzoate (*p*ABA; produced in the chloroplast) in the mitochondria (Basset et al. 2005; Neuburger et al. 1996), and likely as in bacteria, is the site of toxic action for sulfonamide antibiotics in plants. Dihydrofolate synthase, which has similarly been identified in plant mitochondria, catalyzes the reduction of dihydrofolate to tetrahydrofolate (see Fig. 1) as in their bacterial counterparts (Basset et al. 2005).

Presently, there are no data available for the corresponding proteins in algae or fungi (Rébeillé et al. 1997); however, the folate biosynthetic pathway has been identified in apicoplasts of *Plasmodium falciparum*, and studies have investigated the efficacy of sulfonamide drugs on dihydropteroate synthase and genetic sequence alterations leading to sulfonamide resistance (Brooks et al. 1994; Khalil et al. 2003; Wang et al. 1997). Resistance to trimethoprim has also been documented in *P. falciparum* through alterations in dihydrofolate reductsase alleles (Iyer et al. 2001; Khalil et al. 2003).

D. Chloroplastic Fatty Acid Biosynthetic Pathway as a Target for Triclosan

Fatty acid synthesis occurs as iterative elongations of acyl chains utilizing the 2-carbon donor malonyl coenzyme A (CoA), where fatty acid synthase (FAS) is the principal enzymatic unit of this process, and in bacterial systems separate proteins constitute the several enzyme activities of FAS (Waller et al. 2003). Although the reactions catalyzed by FAS are essentially the same for all organisms, two different types are found in nature: type I FAS, found in animals and yeast, consists of a single large multifunctional enzyme complex, and type II FAS, found in plants and most bacteria, consists of multiple proteins with separate enzymatic functions (Harwood 1996). The type I FAS complex is therefore considered to have resulted from gene fusion events converging as a single protein (Smith 1994).

Fatty acid biosynthesis in plants takes place within plastids, and it is thus not surprising that fatty acid metabolism in plants utilizes the bacterial type II system because the plastid is derived from a cyanobacterial endosymbiont (Harwood 1996; Ryall et al. 2003). Long-chain fatty acids are assembled two carbons at a time, and acetyl-CoA carboxylase (ACCase) catalyzes the committed step, the conversion of acetyl-CoA to malonyl-CoA (Harwood 1996). Malonyl-CoA is subsequently transferred to an acyl carrier protein (ACP) by malonyl-CoA:ACP transacylase (FabD) (Harwood 1996). Malonyl-ACP and acetyl-CoA are then condensed, reduced, dehydrated, and reduced once again, yielding an acyl-ACP (originally with four carbon units) (Ryall et al. 2003). The first condensation reaction is catalyzed by β-ketoacyl-ACP synthase III (FabH), subsequent condensation reactions are catalyzed by β-ketoacyl-ACP synthase I and II (FabB and/or FabF), the first reduction step is catalyzed by β-ketoacyl-ACP reductase (FabG), dehydration is catalyzed by β-hydroxyacyl-ACP dehydratase (FabZ), and the final reduction is catalyzed by enoyl-ACP reductase (FabI) (Waller et al. 2003). The chain is elongated two carbon units at a time by condensing another malonyl-ACP with the acyl-ACP and repeating the reaction cycle (Ryall et al. 2003). Pharmacologically, FabI is a major point of regulation for bacterial and plastid fatty acid biosynthesis and is a drug target for triclosan (see Fig. 1) (Ryall et al. 2003). Triclosan is a chlorophenol antibiotic often used in personal care products such as antibacterial soaps and toothpastes that inhibits fatty acid synthesis in bacteria by dramatically increasing the affinity of FabI for NAD+; this creates an FabI-NAD+-triclosan complex where triclosan forms hydrogen bonds and hydrophobic interactions with both the protein and the NAD+ cofactor resulting in a stable ternary complex, with the drug binding at the enoyl substrate site (Heath et al. 1999). Given the high degree of similarity between bacterial and plant (plastid) fatty acid biosynthesis, which suggests evolutionary conservation of this pathway, and the inhibitory effects of triclosan on bacterial FabI, plants are likely susceptible to this compound via the same receptor.

E. Mevalonic Acid Pathway in Plants as a Target for Statin Lipid Regulators

Statins are a class of pharmaceuticals prescribed to lower total cholesterol and low-density lipoprotein cholesterol (LDLc) (Sweetman 2002). These compounds are highly potent competitive inhibitors of 3-hydroxy-3-methylglutaryl coenzyme-A reductase (HMGR) (Alberts et al. 1980), the rate-limiting enzyme in cholesterol biosynthesis. Similarly, HMGR is also present in higher plants as the key enzyme regulating the mevalonic acid pathway (see Fig. 1) of isoprenoid biosynthesis (Disch et al. 1998a) and, as in humans, statins are highly specific inhibitors of this enzyme (Bach and Lichtenthaler 1982a). In plants, there are two discrete pathways for synthesizing the universal precursors to isoprenoid biosynthesis [isopentyl diphosphate (IPP) and its isomer dimethylallyl phosphate (DMAPP)]: a statin-susceptible cytosolic (mevalonate, MVA) pathway of eukaryotic origin and an unsusceptible chloroplastic (2-C-methyl-D-erythritol 4-phosphate, (MEP) pathway of prokaryotic origin (Lichtenthaler et al. 1997a,b; Rohmer 1999).

Over the course of evolution, plants have maintained the eukaryotic MVA pathway in the cytosol but also acquired the more recently discovered prokaryotic MEP pathway (Hemmerlin et al. 2003). Cytosolic formation of IPP proceeds through the intermediate mevalonate from acetyl-CoA, which, under normal physiological conditions, is responsible for the formation of sterols and ubiquinone (Disch et al. 1998a; Lichtenthaler et al. 1997a). Chloroplastic formation of IPP involves a condensation of pyruvate and glyceraldehyde-3-phosphate via a 1-deoxy-D-xylulose 5-phosphate intermediate and is used to form isoprene, carotenoids, abscisic acid, and the side chains of chlorophylls and plastoquinone (Arigoni et al. 1997; Disch et al. 1998a; Lichtenthaler et al. 1997a; Schwender et al. 1996, 1997).

Lange et al. (2000) suggested that the evolutionary history of the enzymes involved in both routes of synthesis could be as follows. The phylogenetic distribution of genes across genomes suggests that the MVA pathway originated from the Archaebacteria, that the MEP pathway derived from the Eubacteria, and that eukaryotes have inherited their genes for IPP biosynthesis from prokaryotes. Green algae (chlorophytes) including *Scenedesmus obliquus*, *C. reinhardtii*, and *Chlorella fusca* use the MEP pathway exclusively, whereas the red alga (rhodophyte) *Cyanidium caldarum* and the diatom (chrysophyte) *Ochromonas danica* possess both pathways, as in higher plants (Disch et al. 1998b). In contrast, the protist (euglenophyte) *Euglena gracilis* exclusively uses the MVA pathway for the synthesis of all its isoprenoids (Disch et al. 1998b). Statins are known inhibitors of sterol biosynthesis in plants, where mevinolin (lovastatin) has been used for decades to study isoprenoid biosynthesis (Bach and Lichtenthaler 1983; Disch et al. 1998a; Hemmerlin et al. 2003).

VI. Toxicity of Pharmaceuticals in Plants

Although the issue of pharmaceuticals in the environment and a corresponding interest in potential toxicological implications has surfaced largely over the last decade, reviews of antibiotic toxicity to plants have been published as far back as the 1950s, indicating that most antibiotics that had been examined at the time showed phytotoxicity to some extent (Brian 1957). Although early studies documented the inhibitory effect of a number of antibiotics in higher plants, the assessments were focused on implications for crop production and not on the ecotoxicological considerations of the discharge of these compounds into the environment and their nontarget toxicity. The following section focuses explicitly on ecotoxicological considerations. All toxicological data pertaining to the following sections are summarized in Tables 1 and 2 for aquatic higher plants and algae, respectively.

A. Antibiotics: Chloroplastic Replication Inhibitors

Fluoroquinolones (DNA Gyrase Inhibitors)

Shortly after its development in the early 1960s, nalidixic acid was screened for antichloplastic activity in the facultative chloroplasts of *Euglena gracilis*, causing a loss of green colony-forming ability at 50 μg/L, and was suspected to have a mode of action analogous to that in bacteria (Lyman et al. 1967). In bacteria, fluoroquinolones inhibit DNA gyrases (DNA topoisomerase) involved in the negative supercoiling of DNA (Champoux 2001; Divo et al. 1988). Among the antibiotics tested for phytotoxicity, quinolones have generally proven to be the most potent class. Ciprofloxacin, lomefloxacin, ofloxacin, levofloxacin, and norfloxacin have all been screened for phytotoxicity in *L. gibba*, with 7-d fresh weight EC_{50} values ranging from 97 to 913 μg/L

Table 1. Toxicity of pharmaceutical compounds to aquatic higher plants, reported as the effective concentration required to cause a decrease in the endpoint of interest by 50% (EC50s). Associated confidence intervals (CIs) or standard errors (SEs) are given where available.

Plant Species	Compound	Class (Antibiotics)	Endpoint	EC_{50} (95% CI) or ± SE Reported in (μg/L)	Test Type	Source
Lemna gibba	Lomefloxacin	Quinolone	Wet Mass	97 (57, 136)	Static Renewal 7d	(Brain et al. 2004b)
Lemna minor	Lomefloxacin	Quinolone	Growth Rate	106 (45–167)	Static Renewal 7d	(Robinson et al. 2005)
Lemna gibba	Levofloxacin	Quinolone	Wet Mass	185 (120, 251)	Static Renewal 7d	(Brain et al. 2004b)
Lemna minor	Levofloxacin	Quinolone	Growth Rate	51 (8.6, 94)	Static Renewal 7d	(Robinson et al. 2005)
Lemna gibba	Ciprofloxacin	Quinolone	Wet Mass	698 (477, 919)	Static Renewal 7d	(Brain et al. 2004b)
Lemna minor	Ciprofloxacin	Quinolone	Growth Rate	203 (41, 364)	Static Renewal 7d	(Robinson et al. 2005)
Lemna gibba	Ofloxacin	Quinolone	Wet Mass	532 (378, 686)	Static Renewal 7d	(Brain et al. 2004b)
Lemna minor	Ofloxacin	Quinolone	Growth Rate	126 (52, 201)	Static Renewal 7d	(Robinson et al. 2005)
Lemna gibba	Norfloxacin	Quinolone	Wet Mass	913 (630, 1,195)	Static Renewal 7d	(Brain et al. 2004b)
Lemna minor	Clinafloxacin	Quinolone	Growth Rate	62 (21–103)	Static Renewal 7d	(Robinson et al. 2005)
Lemna minor	Enrofloxacin	Quinolone	Growth Rate	114 (84, 143)	Static Renewal 7d	(Robinson et al. 2005)
Lemna minor	Flumequine	Quinolone	Growth Rate	2,470 (1,650, 3,300)	Static Renewal 7d	(Robinson et al. 2005)
Lemna gibba	Sulfamethoxazole	Sulfonamide	Wet Mass	81 (52, 109)	Static Renewal 7d	(Brain et al. 2004b)
Lemna gibba	Sulfadimethoxine	Sulfonamide	Wet Mass	248 (178, 318)	Static Renewal 7d	(Brain et al. 2004b)
Lemna gibba	Sulfamethazine	Sulfonamide	Wet Mass	1,277 (823, 1,730)	Static Renewal 7d	(Brain et al. 2004b)
Lemna minor	Sulfachlorpyridazine	Sulfonamide	NA	2,330 (1,400–3,900)	Static 7d	(Pro et al. 2003)
Lemna gibba	Trimethoprim	Diaminopyrimidine	Wet Mass	>1,000 (NR)	Static Renewal 7d	(Brain et al. 2004b)
Lemna gibba	Lincomycin	Lincosamide	Wet Mass	>1,000 (NR)	Static Renewal 7d	(Brain et al. 2004b)
Lemna gibba	Erythromycin	Macrolide	Wet Mass	>1,000 (NR)	Static Renewal 7d	(Brain et al. 2004b)
Lemna minor	Erythromycin	Macrolide	Frond Number	5,620 (NA)	Static Renewal 7d	(Pomati et al. 2004)
Lemna gibba	Roxithromycin	Macrolide	Wet Mass	>1,000 (NR)	Static Renewal 7d	(Brain et al. 2004b)
Lemna gibba	Tylosin	Macrolide	Wet Mass	>1,000 (NR)	Static Renewal 7d	(Brain et al. 2004b)
Lemna gibba	Streptomycin	Aminoglycoside	Wet Mass	>1,000 (NR)	Static Renewal 7d	(Brain et al. 2004b)
Lemna gibba	Neomycin	Aminoglycoside	Wet Mass	>1,000 (NR)	Static Renewal 7d	(Brain et al. 2004b)
Lemna gibba	Chlortetracycline	Tetracycline	Wet Mass	219 (156, 282)	Static Renewal 7d	(Brain et al. 2004b)
Lemna gibba	Doxycycline	Tetracycline	Wet Mass	316 (200, 432)	Static Renewal 7d	(Brain et al. 2004b)
Lemna gibba	Tetracycline	Tetracycline	Wet Mass	723 (480, 966)	Static Renewal 7d	(Brain et al. 2004b)
Lemna minor	Tetracycline	Tetracycline	Frond Number	1,060 (NA)	Static Renewal 7d	(Pomati et al. 2004)
Lemna gibba	Oxytetracycline	Tetracycline	Wet Mass	1,010 (NR)	Static Renewal 7d	(Brain et al. 2004b)
Lemna minor	Oxytetracycline	Tetracycline	(NA)	4,920 (3,600–6,800)	Static 7d	(Pro et al. 2003)
Lemna gibba	Cephalexin	Carbacephem	Wet Mass	>1,000 (NR)	Static Renewal 7d	(Brain et al. 2004b)

Table 1. (*cont.*)

Plant Species	Compound	Class (Antibiotics)	Endpoint	EC_{50} (95% CI) or ± SE Reported in (μg/L)	Test Type	Source
Lemna gibba	Monensin	Ionophore	Wet Mass	998 (955, 1,042)	Static Renewal 7d	(Brain et al. 2004b)
Lemna gibba	Amoxacillin	β-Lactam	Wet Mass	>1,000 (NR)	Static Renewal 7d	(Brain et al. 2004b)
Lemna gibba	Triclosan	Chlorophenol	Frond Number	>62.5	Static 7d	(Orvos et al. 2002)
Plant Species	**Compound**	**Class (Other)**	**Endpoint**	**EC_{50} (95% CI) or ± SE Reported in (μg/L)**	**Test Type**	**Source**
Lemna gibba	Atorvastatin	Statin lipid regulator	Wet Mass	135 ± 27.1	Static Renewal 7d	(Brain et al. 2006)
Lemna gibba	Lovastatin	Statin lipid regulator	Wet Mass	106 ± 8.27	Static Renewal 7d	(Brain et al. 2006)
Lemna gibba	Ibuprofen	NSAID	Wet Mass	>1,000 (NR)	Static Renewal 7d	(Brain et al. 2004b)
Lemna minor	Ibuprofen	NSAID	Growth rate	22,000 (NA)	Static 7d	(Cleuvers 2003)
Lemna minor	Ibuprofen	NSAID	Frond Number	4,010 (NA)	Static Renewal 7d	(Pomati et al. 2004)
Lemna minor	Diclofenac	NSAID	Growth rate	7,500 (NA)	Static 7d	(Cleuvers 2003)
Lemna minor	Naproxen	NSAID	Growth rate	24,200 (NA)	Static 7d	(Cleuvers 2003)
Lemna gibba	Acetaminophen	Analgesic	Wet mass	>1,000 (NR)	Static Renewal 7d	(Brain et al. 2004a)
Lemna gibba	Fluoxetine	SSRI	Wet mass	>1,000 (NR)	Static Renewal 7d	(Brain et al. 2004b)
Lemna gibba	Sertraline	SSRI	Wet mass	>1,000 (NR)	Static Renewal 7d	(Brain et al. 2004a)
Lemna gibba	Cotinine	Nicotine	Wet mass	>1,000 (NR)	Static Renewal 7d	(Brain et al. 2004b)
Lemna gibba	Caffeine	Stimulant	Wet mass	>1,000 (NR)	Static Renewal 7d	(Brain et al. 2004a)
Lemna gibba	Carbamazepine	Anti-Epileptic	Wet mass	>1,000 (NR)	Static Renewal 7d	(Brain et al. 2004a)
Lemna minor	Carbamazepine	Anti-Epileptic	Growth rate	25,500 (NA)	Static 7d	(Cleuvers 2003)
Lemna minor	Clofibrinic acid	Lipid regulator	Growth rate	12,500 (NA)	Static 7d	(Cleuvers 2003)
Lemna minor	Captopril	Vasodilator	Growth rate	25,000 (NA)	Static 7d	(Cleuvers 2003)
Lemna minor	Metformin	Antihyperglycaemic	Growth rate	110,000 (NA)	Static 7d	(Cleuvers 2003)
Lemna minor	Propranolol	B-blocker	Growth rate	114,000 (NA)	Static 7d	(Cleuvers 2003)
Lemna minor	Metoprolol	B-blocker	Growth rate	>320,000 (NA)	Static 7d	(Cleuvers 2003)

The values reported are in μg/L of respective pharmaceutical.
The acronym NR refers to "not reported" with respect to the 95% confidence interval or standard error.
The acronym NA refers to "not available" due to lack repordence of a concentration-response or convergence.
The acronyms NSAID and SSRI refer to non-steroidal anti-inflammatory drug and selective serotonin reuptake inhibitor, respectively.

Table 2. Toxicity of pharmaceutical compounds to algae, reported as the effective concentrations required to cause a decrease in the endpoint of interest by 50% (EC50s). Associated confidence intervals (CIs) or standard errors (SEs) are given where available.

Algal Species	Algal Type	Compound	Class (Antibiotics)	Endpoint	EC_{50} (95% CI) or ± SE Reported in (μg/L)	Test Type	Source
Microcystis aeruginosa	Blue-green	Olaquindox	Quinoxaline	Chlorophyll	5,100 (4,500–5,600)	Static 7d	(Halling-Sørensen 2000)
Selenastrum capricornutum	Green	Olaquindox	Quinoxaline	Chlorophyll	40,000 (26,000–58,000)	Static 3d	(Halling-Sørensen 2000)
Microcystis aeruginosa	Blue-green	Streptomycin	Aminoglycoside	Chlorophyll	7 (6–8)	Static 7d	(Halling-Sørensen 2000)
Selenastrum capricornutum	Green	Streptomycin	Aminoglycoside	Chlorophyll	133 (34–357)	Static 3d	(Halling-Sørensen 2000)
Microcystis aeruginosa	Blue-green	Tiamulin	Pleuromutilin	Chlorophyll	3 (2–4)	Static 7d	(Halling-Sørensen 2000)
Selenastrum capricornutum	Green	Tiamulin	Pleuromutilin	Chlorophyll	165 (79–285)	Static 3d	(Halling-Sørensen 2000)
Microcystis aeruginosa	Blue-green	Spiramycin	Macrolide	Chlorophyll	5 (1–18)	Static 7d	(Halling-Sørensen 2000)
Selenastrum capricornutum	Green	Spiramycin	Macrolide	Chlorophyll	2,300 (1,300–4,000)	Static 3d	(Halling-Sørensen 2000)
Microcystis aeruginosa	Blue-green	Tylosin	Macrolide	Chlorophyll	34 (24–48)	Static 7d	(Halling-Sørensen 2000)
Selenastrum capricornutum	Green	Tylosin	Macrolide	Chlorophyll	1,380 (970–1,960)	Static 3d	(Halling-Sørensen 2000)
Selenastrum capricornutum	Green	Erythromycin	Macrolide	Number of cells	20 (16–26)	Static 3d	(Isidori et al. 2005)
Selenastrum capricornutum	Green	Lincomycin	Macrolide	Number of cells	70 (50–100)	Static 3d	(Isidori et al. 2005)
Selenastrum capricornutum	Green	Clarithromycin	Macrolide	Number of cells	2 (1.9–2.8)	Static 3d	(Isidori et al. 2005)
Microcystis aeruginosa	Blue-green	Benzylpenicillin	β-Lactam	Chlorophyll	6 (4–12)	Static 7d	(Halling-Sørensen 2000)
Selenastrum capricornutum[a]	Green	Benzylpenicillin	β-Lactam	Chlorophyll	>100,000 (NR)	Static 3d	(Halling-Sørensen 2000)
Microcystis aeruginosa	Blue-green	Amoxacillin	β-Lactam	Chlorophyll	3.7 (NR)	Static 7d	(Holten Lützhøft et al. 1999)
Selenastrum capricornutum	Green	Amoxacillin	β-Lactam	Chlorophyll	250,000 (NOEC)	Static 3d	(Holten Lützhøft et al. 1999)
Selenastrum capricornutum	Green	Amoxacillin	β-Lactam	Chlorophyll	5,000 (1,600–16,000)	Static 3d	(Holten Lützhøft et al. 1999)
Selenastrum capricornutum	Green	Flumequine	Quinolone	Chlorophyll	5,000 (4,800–5,200)	Static 3d	(Robinson et al. 2005)
Microcystis aeruginosa	Blue-green	Flumequine	Quinolone	Chlorophyll	1,960 (1,760–2,160)	Static 5d	(Robinson et al. 2005)
Microcystis aeruginosa	Blue-green	Flumequine	Quinolone	Chlorophyll	159 (66–382)	Static 7d	(Holten Lützhøft et al. 1999)
Microcystis aeruginosa	Blue-green	Oxolinic acid	Quinolone	Chlorophyll	180 (NR)	Static 7d	(Holten Lützhøft et al. 1999)
Selenastrum capricornutum	Green	Oxolinic acid	Quinolone	Chlorophyll	16,000 (9,100–29,000)	Static 3d	(Holten Lützhøft et al. 1999)
Selenastrum capricornutum	Green	Levofloxacin	Quinolone	Chlorophyll	7,400 (6,400–8,400)	Static 3d	(Robinson et al. 2005)
Microcystis aeruginosa	Blue-green	Levofloxacin	Quinolone	Chlorophyll	7.9 (6.4–9.4)	Static 5d	(Robinson et al. 2005)
Selenastrum capricornutum	Green	Ciprofloxacin	Quinolone	Chlorophyll	18,700 (16,200–21,200)	Static 3d	(Robinson et al. 2005)
Microcystis aeruginosa	Blue-green	Ciprofloxacin	Quinolone	Chlorophyll	17 (14–20)	Static 5d	(Robinson et al. 2005)
Selenastrum capricornutum	Green	Enrofloxacin	Quinolone	Chlorophyll	3,100 (2,600–3,600)	Static 3d	(Robinson et al. 2005)

Table 2. (*cont.*)

Algal Species	Algal Type	Compound	Class (Antibiotics)	Endpoint	EC_{50} (95% CI) or ± SE Reported in (μg/L)	Test Type	Source
Microcystis aeruginosa	Blue-green	Enrofloxacin	Quinolone	Chlorophyll	49 (41–56)	Static 5d	(Robinson et al. 2005)
Selenastrum capricornutum	Green	Clinafloxacin	Quinolone	Chlorophyll	1,100 (930–1,300)	Static 3d	(Robinson et al. 2005)
Microcystis aeruginosa	Blue-green	Clinafloxacin	Quinolone	Chlorophyll	103 (86–120)	Static 5d	(Robinson et al. 2005)
Selenastrum capricornutum	Green	Lomefloxacin	Quinolone	Chlorophyll	22,700 (19,900–25,500)	Static 3d	(Robinson et al. 2005)
Microcystis aeruginosa	Blue-green	Lomefloxacin	Quinolone	Chlorophyll	186 (172–200)	Static 5d	(Robinson et al. 2005)
Selenastrum capricornutum	Green	Ofloxacin	Quinolone	Chlorophyll	12,100 (10,400–13,700)	Static 3d	(Robinson et al. 2005)
Microcystis aeruginosa	Blue-green	Ofloxacin	Quinolone	Chlorophyll	21 (18–24)	Static 5d	(Robinson et al. 2005)
Selenastrum capricornutum	Green	Ofloxacin	Quinolone	Number of cells	4,740 (NR)	Static 4d	(Ferrari et al. 2004)
Synechococcus leopoliensis	Blue-green	Ofloxacin	Quinolone	Number of cells	16 (NR)	Static 4d	(Ferrari et al. 2004)
Selenastrum capricornutum	Green	Ofloxacin	Quinolone	Number of cells	1,440 (1,080–1,800)	Static 3d	(Isidori et al. 2005)
Microcystis aeruginosa	Blue-green	Sarafloxacin	Quinolone	Chlorophyll	15[b] (9–23)	Static 7d	(Holten Lützhøft et al. 1999)
Selenastrum capricornutum	Green	Sarafloxacin	Quinolone	Chlorophyll	16,000 (9,800–25,000)	Static 3d	(Holten Lützhøft et al. 1999)
Microcystis aeruginosa	Blue-green	Chlortetracycline	Tetracycline	Chlorophyll	50 (30–100)	Static 7d	(Halling-Sørensen 2000)
Selenastrum capricornutum	Green	Chlortetracycline	Tetracycline	Chlorophyll	3,100 (1,700–5,200)	Static 3d	(Halling-Sørensen 2000)
Microcystis aeruginosa	Blue-green	Tetracycline	Tetracycline	Chlorophyll	9 (8–10)	Static 7d	(Halling-Sørensen 2000)
Selenastrum capricornutum	Green	Tetracycline	Tetracycline	Chlorophyll	2,200 (1,500–2,900)	Static 3d	(Halling-Sørensen 2000)
Microcystis aeruginosa	Blue-green	Oxytetracycline	Tetracycline	Chlorophyll	207 (175–246)	Static 7d	(Holten Lützhøft et al. 1999)
Selenastrum capricornutum	Green	Oxytetracycline	Tetracycline	Chlorophyll	4,500 (2,300–8,600)	Static 3d	(Holten Lützhøft et al. 1999)
Selenastrum capricornutum	Green	Oxytetracycline	Tetracycline	Number of cells	170 (110–250)	Static 3d	(Isidori et al. 2005)
Microcystis aeruginosa	Blue-green	Sulfadiazine	Sulfonamide	Chlorophyll	135 (82–223)	Static 7d	(Holten Lützhøft et al. 1999)
Selenastrum capricornutum	Green	Sulfadiazine	Sulfonamide	Chlorophyll	7,800 (4,500–14,000)	Static 3d	(Holten Lützhøft et al. 1999)
Selenastrum capricornutum	Green	Sulfamethoxazole	Sulfonamide	Number of cells	146 (NR)	Static 4d	(Ferrari et al. 2004)
Selenastrum capricornutum	Green	Sulfamethoxazole	Sulfonamide	Number of cells	520 (360–740)	Static 3d	(Isidori et al. 2005)
Synechococcus leopoliensis	Blue-green	Sulfamethoxazole	Sulfonamide	Number of cells	26.8 (NR)	Static 4d	(Ferrari et al. 2004)
Microcystis aeruginosa	Blue-green	Trimethoprim	Diaminopyrimidine	Chlorophyll	112,000 (100,000–126,000)	Static 7d	(Holten Lützhøft et al. 1999)
Selenastrum capricornutum	Green	Trimethoprim	Diaminopyrimidine	Chlorophyll	130,000 (81,000–211,000)	Static 3d	(Holten Lützhøft et al. 1999)
Selenastrum capricornutum	Green	Triclosan	Chlorophenol	Number of cells	4.46 (2.06–9.66)	NR	(Orvos et al. 2002)
Scenedesmus subspicatus	Green	Triclosan	Chlorophenol	Number of cells	0.7 (0.7–0.8)	Static 3d	(Orvos et al. 2002)

Table 2. (*cont.*)

Algal Species	Algal Type	Compound	Class (Antibiotics)	Endpoint	EC_{50} (95% CI) or ± SE Reported in (μg/L)	Test Type	Source
Scenedesmus subspicatus	Green	Triclosan	Chlorophenol	Number of cells	1.4 (1.4–1.5)	Static 4d	(Orvos et al. 2002)
Anabaena flos-aquae	Blue-green	Triclosan	Chlorophenol	Number of cells	0.97 (0.72–1.30)	4d	(Orvos et al. 2002)
Navicula pelliculosa	Diatom	Triclosan	Chlorophenol	NR	19.1 (15.3–23.8)	NR	(Orvos et al. 2002)

Algal Species	Algal Type	Compound	Class (Other)	Endpoint	EC_{50} (95% CI) or ± SE Reported in (μg/L)	Test Type	Source
Chlorella sp.	Green	Metronidazole	Nitroimidazole	Chlorophyll	12,500 (3,440–45,300) 38,800 (27,700–54,400) 45,100 (28,700–70,900)	Static 3d	(Lanzky and Halling-Sørensen 1997)
Selenastrum Capricornutum	Green	Metronidazole	Nitroimidazole	Chlorophyll	39,100 (32,300–47,400) 39,000 (32,300–47,400)	Static 3d	(Lanzky and Halling-Sørensen 1997)
Scenedesmus subspicatus[c]	Green	Diclofenac	NSAID	Chlorophyll	71,900 (65,500–79,100)	Static 3d	(Cleuvers 2004)
Scenedesmus subspicatus	Green	Diclofenac	NSAID	Chlorophyll	72,000 (NR)	Static 3d	(Cleuvers 2003)
Selenastrum capricornutum	Green	Diclofenac	NSAID	Number of cells	10,000 (NOEC)	Static 4d	(Ferrari et al. 2003)
Selenastrum capricornutum	Green	Diclofenac	NSAID	Number of cells	16,300 (NR)	Static 4d	(Ferrari et al. 2004)
Synechococcus leopoliensis	Blue-green	Diclofenac	NSAID	Number of cells	14,500 (NR)	Static 4d	(Ferrari et al. 2004)
Scenedesmus subspicatus	Green	Ibuprofen	NSAID	Chlorophyll	342,200 (242,400–471,500)	Static 3d	(Cleuvers 2004)
Scenedesmus subspicatus	Green	Ibuprofen	NSAID	Chlorophyll	315,000 (NR)	Static 3d	(Cleuvers 2003)
Scenedesmus subspicatus	Green	Naproxen	NSAID	Chlorophyll	625,500 (NR)	Static 3d	(Cleuvers 2004)
Scenedesmus subspicatus	Green	Acetylsalicylic acid	NSAID	Chlorophyll	106,700 (104,400–114,300)	Static 3d	(Cleuvers 2004)
Scenedesmus subspicatus	Green	Salicylic acid	NSAID	Number of cells	>100,000 (NR)	Static 3d	(Henschel et al. 1997)
Scenedesmus subspicatus	Green	Paracetamol	Analgesic	Number of cells	134,000 (NR)	Static 3d	(Henschel et al. 1997)
Selenastrum capricornutum	Green	Carbamazepine	Anti-epileptic	Number of cells	>100,000 (NOEC)	Static 4d	(Ferrari et al. 2003)
Scenedesmus subspicatus	Green	Carbamazepine	Anti-epileptic	Chlorophyll	74,000 (NR)	Static 3d	(Cleuvers 2003)
Selenastrum capricornutum	Green	Carbamazepine	Anti-epileptic	Number of cells	>100,000 (NR)	Static 4d	(Ferrari et al. 2004)
Synechococcus leopoliensis	Blue-green	Carbamazepine	Anti-epileptic	Number of cells	33,600 (NR)	Static 4d	(Ferrari et al. 2004)
Selenastrum capricornutum	Green	Clofibric acid	Lipid regulator	Number of cells	75,000 (NOEC)	Static 4d	(Ferrari et al. 2003)
Selenastrum capricornutum	Green	Clofibric acid	Lipid regulator	Number of cells	94,000 (NR)	Static 4d	(Ferrari et al. 2004)

Table 2. (*cont.*)

Algal Species	Algal Type	Compound	Class (Other)	Endpoint	EC_{50} (95% CI) or ± SE Reported in (µg/L)	Test Type	Source
Synechococcus leopoliensis	Blue-green	Clofibric acid	Lipid regulator	Number of cells	40,200 (NR)	Static 4d	(Ferrari et al. 2004)
Scenedesmus subspicatus	Green	Clofibrinic acid	Lipid regulator	Number of cells	89,000 (NR)	Static 3d	(Henschel et al. 1997)
Scenedesmus subspicatus	Green	Clofibrinic acid	Lipid regulator	Chlorophyll	115,000 (NR)	Static 3d	(Cleuvers 2003)
Scenedesmus subspicatus	Green	Methotrexate	Antimetabolite	Number of cells	260,000 (NR)	Static 3d	(Henschel et al. 1997)
Selenastrum capricornutum	Green	Fluoxetine	SSRI	Number of cells	44.99 ± 1.76	Static 4d	(Johnson et al. 2006)
Selenastrum capricornutum	Green	Fluoxetine	SSRI	Number of cells	24 (NR)	Static 4d	(Brooks et al. 2003)
Scenedesmus acutus	Green	Fluoxetine	SSRI	Number of cells	91.23 ± 2.74	Static 4d	(Johnson et al. 2006)
Scenedesmus quadricauda	Green	Fluoxetine	SSRI	Number of cells	212.98 ± 16.13	Static 4d	(Johnson et al. 2006)
Chlorella vulgaris	Green	Fluoxetine	SSRI	Number of cells	4,339.25 ± 446.09	Static 4d	(Johnson et al. 2006)
Selenastrum capricornutum	Green	Fluvoxamine	SSRI	Number of cells	4,002.88 ± 142.52	Static 4d	(Johnson et al. 2006)
Scenedesmus acutus	Green	Fluvoxamine	SSRI	Number of cells	3,620.24 ± 134.96	Static 4d	(Johnson et al. 2006)
Scenedesmus quadricauda	Green	Fluvoxamine	SSRI	Number of cells	3,563.34 ± 118.94	Static 4d	(Johnson et al. 2006)
Chlorella vulgaris	Green	Fluvoxamine	SSRI	Number of cells	10,208.47 ± 379.24	Static 4d	(Johnson et al. 2006)
Selenastrum capricornutum	Green	Sertraline	SSRI	Number of cells	12.10 ± 1.00	Static 4d	(Johnson et al. 2006)
Scenedesmus acutus	Green	Sertraline	SSRI	Number of cells	98.92 ± 6.74	Static 4d	(Johnson et al. 2006)
Scenedesmus quadricauda	Green	Sertraline	SSRI	Number of cells	317.02 ± 21.46	Static 4d	(Johnson et al. 2006)
Chlorella vulgaris	Green	Sertraline	SSRI	Number of cells	763.66 ± 25.42	Static 4d	(Johnson et al. 2006)
Scenedesmus subspicatus	Green	Captopril	Vasodilator	Chlorophyll	168,000 (NR)	Static 3d	(Cleuvers 2003)
Scenedesmus subspicatus	Green	Metformin	Antihyperglycaemic	Chlorophyll	>320,000 (NR)	Static 3d	(Cleuvers 2003)
Scenedesmus subspicatus	Green	Propanolol	B-blocker	Chlorophyll	5,800 (NR)	Static 3d	(Cleuvers 2003)
Selenastrum capricornutum	Green	Propranolol	B-blocker	Number of cells	7,400 (NR)	Static 4d	(Ferrari et al. 2004)
Synechococcus leopoliensis	Blue-green	Propranolol	B-blocker	Number of cells	668 (NR)	Static 4d	(Ferrari et al. 2004)
Scenedesmus subspicatus	Green	Metoprolol	B-blocker	Chlorophyll	7,300 (NR)	Static 3d	(Cleuvers 2003)

The values reported are in µg/L of respective pharmaceutical.
The acronym NR refers to "not reported" with respect to the 95% confidence interval or standard error.
The acronym NA refers to "not available" due to lack repordence of a concentration-response or convergence.
The acronyms NSAID and SSRI refer to non-steroidal anti-inflammatory drug and selective serotonin reuptake inhibitor, respectively.
The acronym NOEC refers to no-observable effect concentration.
[a]Recently renamed *Pseudokirchneriella subcapitata*.
[b]Extrapolated outside the range of concentrations tested.
[c]Recently renamed *Desmodesmus subspicatus*.

(Brain et al. 2004b). Similarly, in *L. minor*, ciprofloxacin, lomefloxacin, ofloxacin, levofloxacin, clinafloxacin, enrofloxacin, and flumequine were all found to be phytotoxic in 7-d tests, with frond EC_{50} values ranging from 53 to 2,470 µg/L (Robinson et al. 2005), showing good agreement with *L. gibba*. Not all aquatic plants are highly sensitive to fluorquinolones, however, as flumequine is stimulatory to the aquatic weed *Lythrum salicaria* at concentrations between 50 and 5,000 µg/L, only becoming inhibitory at 100 mg/L (Migliore et al. 2000). Cyanobacteria such as *M. aeruginosa* are comparably sensitive to *Lemna* sp. when exposed to fluoroquinolones, with EC_{50} values for growth ranging from 7.9 to 1,960 µg/L (Robinson et al. 2005). However, green algae such as *S. capricornutum* are up to two orders of magnitude less sensitive than cyanobacteria and *Lemna* sp., with EC_{50} values ranging from 1,100 to 22,700 µg/L for algal growth (Robinson et al. 2005). The reason for this differential sensitivity is unknown, particularly as topoisomerase have been identified in green algae (*C. reinhardtii*) (Thompson and Mosig 1985), just as in higher plants (Wall et al. 2004).

Chloroplastic type II topoisomerases and DNA replication in *C. reinhardtii* are inhibited with treatment of the coumarin novobiocin; however, after a delay of several hours, chloroplast chromosomes initiate a novobiocin-insensitive mode of DNA replication, which results in partial replication of the chloroplast chromosome (Woelfle et al. 1993). It is uncertain whether this phenomenon exists in the chloroplasts of cyanobacteria and higher plants, although differences within compensatory replication strategies could potentially explain this differential sensitivity. It is also worth noting that parasites of the phylum Apicomplexa, which contain facultative chloroplasts suspected to be acquired through secondary endosymbiosis of a green alga, have similar sensitivity to green algae (Divo et al. 1988; Hashimoto and Murakami 1982; Krajcovic et al. 1989; Lyman et al. 1967; Woods et al. 1996). For example, the IC_{50} for *P. falciparum* treated with ciprofloxacin is 86,150 µg/L (Divo et al. 1988) compared to an EC_{50} of 18,700 µg/L for *S. capricornutum* (Robinson et al. 2005), whereas the EC_{50} values for *L. gibba*, *L. minor*, and *M. aeruginosa* are 698, 203, and 17 µg/L, respectively (Brain et al. 2004b; Robinson et al. 2005).

Symptomology for all photosynthetic organisms exposed to DNA gyrase inhibitors is markedly similar, typified by pronounced bleaching of tissues or cultures (Brain et al. 2004b; Krajcovic et al. 1989; Lyman et al. 1967; Robinson et al. 2005). Algal assemblages from environmental samples taken upstream and downstream of wastewater treatment plants (WWTPs) also show a high degree of sensitivity to ciprofloxacin (Wilson et al. 2003). Notable effects of ciprofloxacin on environmental samples taken upstream of a WWTP include a significant reduction of the common diatom *Navicula* at 0.12 µg/L, reductions in the green alga *Chlamydomonas* at 0.12 µg/L and 1.2 µg /L, and of *Sphaerocystis* at 1.2 µg/L ($P < 0.05$) (Wilson et al. 2003). Consistent and dramatic declines in genus richness were found for increasing ciprofloxacin treatment. In comparisons of certain sampling dates, ciprofloxacin was found to cause statistically significant effects for *Synedra* and *Chlamydomonas* between upstream and downstream sampling sites (Wilson et al. 2003).

Olaquindox

Olaquindox, a quinoxaline antibiotic, is a synthetic chemotherapeutic agent that facilitates peptide nucleic acid (PNA)–DNA complex formation in bacteria, acting as an intercalating agent that inhibits DNA synthesis (Møllegaard et al. 2000). This compound shows little toxicity to either blue-green algae or green algae, with respective EC_{50} values of 5,100 and 40,000 µg/L (Halling-Sørensen 2000).

B. Antibiotics: Chloroplastic Translation, Transcription, and Anabolic Process Inhibitors

Tetracyclines

Chlortetracycline and oxytetracycline have been reported to induce chlorosis in higher plants, as far back as the 1950s (Brian 1957). In bacteria, tetracyclines bind irreversibly to the 30S subunit of ribosomes, blocking the binding of aminoacyl transfer to DNA, inhibiting protein synthesis (Sweetman 2002). Early studies with tetracycline antibiotics, their degradation products, derivatives, and some synthetic analogues conducted on several species of duckweeds, including *L. gibba*, and two species of ferns indicated that oxytetracycline, tetracycline, and chlortetracycline were inhibitory at concentrations of 10 or 20 mg/L in 21-d experiments (Nickell and Gordon 1961). However, more recent studies have demonstrated the toxicity of tetracyclines to *Lemna* sp. at much lower concentrations of 219–4,920 µg/L, under different exposure scenarios (Brain et al. 2004b; Pro et al. 2003).

Nickell and Gordon (1961) suggested that the mode of action of tetracyclines is largely through chelating of metal nutrients in growth solutions and less so via direct toxicity. This hypothesis was supported by a recent microcosm investigation of the effects of a mixture of tetracyclines (tetracycline, chlortetracycline, oxytetracycline, and doxycycline) in 12,000-L aquatic microcosms, where *L. gibba* showed no signs of phytotoxicity because nutrients were not limiting in these experimental units (Brain et al. 2005b). The submerged macrophyte *Myriophyllum sibiricum*, on the other hand, experienced concentration dependent decreases in dry mass to 69%, 47%, 30%, and 7% of controls at treatment concentrations of 0.080, 0.218, 0.668, and 2.289 µmol/L, respectively, as a result of corresponding concentration-dependent reductions in light penetration as high as 99.8% at a depth of 70 cm (Brain et al. 2005b).

The lack of response in *Lemna sp.* and the indirect toxic effect observed for *M. sibiricum* indicate that it is unlikely that tetracyclines caused direct toxicity to the translation apparatus of the chloroplast of these species. Both blue-green and green algae show sensitivity to tetracyclines in laboratory studies, with EC_{50} values of 9–207 µg/L for *M. aeruginosa* and 170–4,500 µg/L *S. capricornutum*, respectively, although generally cyanobacteria are more sensitive (Halling-Sørensen 2000; Holten Lützhøft et al. 1999; Isidori et al. 2005). It is not known whether tetracyclines disrupt translation in algae, or whether toxicity is caused as a result of chelation of mineral nutrients, as has been shown for *Lemna* sp. However, because concentrations required to elicit a response in aquatic microcosms are several orders of magnitude higher than those required for laboratory tests (Wilson et al. 2004), the toxicity observed in laboratory studies is likely largely caused by indirect effects such as chelation. Effects in the field are likely caused by indirect decreases in light penetration, as has been observed for *M. sibiricum* (Brain et al. 2005b). Inhibitory concentrations of tetracyclines in *E. gracilis* are also high, generally well above 1 mg/L, with the exception of doxycycline (Woods et al. 1996), suggesting that the mode of toxic action is likely not the pharmacological mode of action in photosynthetic organisms. Alternatively, tetracyclines may experience poor rates of uptake and/or translocation or high rates of metabolism, although this has not been investigated.

Macrolides and Lincosamides

Pharmacologically, macrolide and lincosamide antibiotics interact with the peptidyl tranferase domain of bacterial 23S rRNA, whereas aminoglycosides irreversibly bind to 30S and 50S subunits of ribosomes (Sweetman 2002). In *Lemna*, erythromycin is inhibitory at relatively high concentrations of 1 mg/L with an extrapolated EC_{50} of 5,620 µg/L, although induction of the stress hormone abscisic acid (ABA), was found to be promoted by erythromycin and tetracycline exposure (Pomati et al. 2004). In laboratory assays with *L. gibba*, tylosin was found to cause an increase in wet mass by 25% at 212 µg/L, and similar stimulator effects on both *L. gibba* and *M. spicatum* have been documented in microcosm studies (Brain et al. 2005a). Macrolides also demonstrate little potency to protoplast-derived cells of *Nicotiana plumbaginifolia* (tobacco) (erythromycin is toxic between 30 and 80 mg/L) (Pollock et al. 1983). Although *S. capricornutum* is relatively insensitive to tylosin and spiramycin with EC_{50}s > 1,000 µg/L (Halling-Sørensen 2000), erythromycin, lincomycin, and clarithromycin inhibit growth by 50% at 20, 70, and 2 µg/L, respectively (Isidori et al. 2005). Compared to *S. capricornutum*, *M. aeruginosa* is considerably more sensitive to spiramycin and tylosin with EC_{50} values of 5 and 34 µg/L, respectively (Halling-Sørensen 2000).

Aminoglycosides and Pleuromutilins

Aminoglycosides irreversibly bind to 30S and 50S subunits of bacterial ribosomes (Sweetman 2002). These compounds show similar trends in toxicity as macrolides and lincosamides, with *M. aeruginosa* and *S. capricornutum* demonstrating sensitivity to streptomycin with EC_{50} values (7–133 μg/L) much lower than those found in *L. gibba* (>1,000 μg/L) (Brain et al. 2004b; Halling-Sørensen 2000; Pollock et al. 1983). However, higher plants are much less sensitive; *L. gibba* showed no significant inhibition in growth at concentrations up to 1,000 μg/L (Brain et al. 2004b). Pollock et al. (1983) tested more than 20 antibiotics for toxicity to protoplast-derived cells of *Nicotiana plumbaginifolia* (Tex-Mex tobacco) and found that streptomycin, neomycin, kanamycin, gentamicin, G418, amikacin, and tobramycin were all toxic to varying degrees, although streptomycin was the least toxic, exhibiting effects only at 100,000 μg/L. Toxicity was suggested to be caused by the action of the aminoglycosides on the "prokaryotic-like" ribosomes of chloroplasts and mitochondria (Pollock et al. 1983).

Streptomycin has been found to inhibit the growth of higher plants and cause bleaching of the chloroplast, as well as reduce protein synthesis by about 75% at a concentration of 50,000 μg/L (May et al. 1969). This effect could be reversed by the addition of divalent cations (e.g., Mg^{2+}) and was suggested to be the consequence of mutual competition for sites of uptake (May et al. 1969). Exposure of some strains of *E. gracilis* to high concentrations (100,000 μg/L) of streptomycin leads to fragmentation and loss of chloroplasts with consequent loss of chlorophyll and photosynthetic capabilities, although only when exposed cells are reproducing vigorously (Brian 1957; Montandon et al. 1985). *Cryptosporidium parvum* is sensitive to streptomycin at high concentrations, where 28% and 100% inhibition is observed at 300,000 and 1,000,000 μg/L (Woods et al. 1996). Tiamulin is a pleuromutilin antibiotic that also binds to the 50S subunit of the bacterial ribosome and interacts at the peptidyl transferase center (Högenauer and Ruf 1981). Similar to macrolides, lincosamides, and aminoglycosides, tiamulin is toxic to both green and blue-green algae, with EC_{50} values of 3 and 165 for *M. aeruginosa* and *S. capricornutum*, respectively (Halling-Sørensen 2000), although no data are available for sensitivity comparisons to higher plants.

Among plant species, blue-green and green algae are considerably more sensitive to bacterial protein inhibitors than aquatic higher plants. It is not surprising that algae are generally highly susceptible to bacterial protein inhibitors because proteomic analysis of chloroplastic ribosomes from *C. reinhardtii* indicates a high degree of homology with the bacterial ribosomes of *E. coli* (Yamaguchi et al. 2003). However, similar to green algae, protein identification of the chloroplastic ribosomes of higher plants has also identified a full complement of *E. coli* ribosomal proteins (Yamaguchi and Subramanian 2000, 2003). This observation suggests that because the appropriate receptor is present in both algae and higher plants, the differential sensitivity between these photosynthetic organisms is likely caused by differences in the uptake, translocation, and/or metabolism of these compounds, although this has not been investigated. For members of the phylum

Apicomplexa such as *E. gracilis* and *C. parvum*, antichloroplastic activity of these bacterial protein inhibitors is only induced at concentrations >>1 mg/L (Ebringer 1972; Nicolas 1981; van Pel and Cocito 1973; Woods et al. 1996).

C. Antibiotics: Folate Biosynthetic Inhibitors

Folates are a group of compounds that are vital cofactors for enzymes, mediating one-carbon transfer reactions. Pharmacologically, sulfonamide antibiotics act as structural analogues of the substrate *p*-aminobenzoic acid (PABA), inhibiting the enzyme dihydropteroate synthase in the folate biosynthetic pathway (Sweetman 2002). Trimethoprim, a diaminopyrimidine antibiotic, similarly inhibits the folate biosynthetic pathway, however, by inhibiting the enzyme dihydrofolate reductase (Sweetman 2002). Both of these target enzymes have been identified in the folate biosynthetic pathway of higher plants, which is essentially the same as in bacteria (Basset et al. 2005). In *Lemna* sp. treated with different sulfonamides, EC_{50} values ranged from 81 to 2,330 µg/L (Brain et al. 2005b; Pro et al. 2003); sulfamethoxazole was the most toxic antibiotic of any class tested with respect to fresh weight in *L. gibba*. Both green (EC_{50}: 146–7,800 µg/L) (Ferrari et al. 2004; Holten Lützhøft et al. 1999) and blue-green algae (EC_{50}: 26.8–135 µg/L) (Holten Lützhøft et al. 1999; Isidori et al. 2005) show remarkably similar sensitivity to *L. gibba*. Sulfamethoxazole is consistently the most potent in each case, and provides the only example where blue-green algae, green algae, and higher plants showed sensitivity to the same antibiotic, varying by less than one order of magnitude. *Plasmodium flaciparum* is comparably less sensitive to sulfamethoxazole (mean IC_{50}: 89 mg/L), although sulfadoxine (mean IC_{50}: 63 mg/L) and trimethoprim (mean IC_{50}: 1.5 mg/L) are slightly more potent (Khalil et al. 2003).

Neither green or blue-green algae nor higher plants show any significant sensitivity toward trimethoprim, with corresponding EC50 values of 130,000, 112,000, and >1,000 µg/L, respectively. Experiments with sulfamethoxazole and trimethoprim in *L. gibba* showed no significant difference between exposures to sulfamethoxazole alone and combinatorial exposures to sulfamethoxazole and trimethoprim at equimolar concentrations (Brain 2006, unpublished data). This discrepancy suggests another instance of differential uptake, translocation, and/or metabolism because receptors exist for both sulfamethoxazole and trimethoprim in plants.

D. Antibiotics: Fatty Acid Synthase Inhibitors

Triclosan is an inhibitor of bacterial fatty acid biosynthesis that specifically targets the enzyme enoyl-ACP reductase or FabI (Heath et al. 1999). This enzyme has been identified in plants, which synthesize fatty acids using the same type II FAS pathway as bacteria (Harwood 1996). Green and blue-green algae and diatoms are highly sensitive to triclosan, with EC_{50}s ranging from 0.7 to 19.1 µg/L, although *L. gibba* is less sensitive with EC_{50} values >62.5 µg/L (the highest concentrations tested) (Orvos et al. 2002). Although *Lemna* had an EC_{50} in excess of the highest concentrations tested, Orvos et al. (2002) noted that studies with all plant species tested confirmed the hypothesis that plants are indeed sensitive to the effects of triclosan, although the specific degree of effect was not stated. A more precise estimate of the EC_{50} for *L. gibba* is required before meaningful comparisons can be made with respect to the margin of differential sensitivity between algae and higher plants.

Although inhibition of FabI has not been demonstrated specifically in plants, triclosan has been shown to inhibit a plant-like FabI in *P. falciparum* at IC_{50} values ranging from approximately 150 to 2,000 µg/L (McLeod et al. 2001). Thiolactomycin, a naturally produced antibiotic from *Nocardia sp.*, is a potent inhibitor of FabB in *E. coli* (Furukawa et al. 1993) that has also been shown to be highly toxic to *P. falciparum* (Waller et al. 2003). In the leaves of the higher plant *Avena* (oat), this compound has been shown to inhibit fatty acid synthesis with an EC_{50} of 380 µg/L (Kato et al. 1987). This finding further suggests a high degree of conservation of enzymes involved in type II FAS between bacteria and plants, and coupled with the low EC_{50} values for triclosan to most plant species, indicates that the toxic response in plants is likely receptor mediated and not via narcosis.

Furthermore, algal assemblages from environmental samples taken upstream of a WWTP also show a high degree of sensitivity to triclosan with respect to absolute algal biomass for chroococcalian cyanobacteria and for the green algae *Chlamydomonas* and *Scenedesmus* (Wilson et al. 2003). Notable effects of triclosan include a significant reduction of *Chlamydomonas* at concentrations of 0.015, 0.15, and 1.5 µg/L, a significant reduction of *Sphaerocystis* at 1.5 µg/L ($P < 0.05$), and a significant reduction of chroococcalian cyanobacteria at the concentration of 0.15 µg/L ($P < 0.05$) (Wilson et al. 2003). Increasing the environmental concentration of triclosan resulted in a consistent and dramatic decline in genus richness (Wilson et al. 2003).

E. Other Antibiotics

β-lactam antibiotics, which include penicillin, largely considered as the genesis of antibiotic therapy, inhibit the formation of peptidoglycan cross-links in the bacterial cell wall by binding to the enzyme (transpeptidase) that links the peptidoglycan molecules (Sweetman 2002). Blue-green algae, such as *M. aeruginosa,* are highly sensitive to both benzylpenicillin and amoxicillin, with respective EC_{50} values of 6 and 3.7 µg/L, although green algae and higher plants are comparatively insensitive to β-lactams, with EC_{50} values generally well above 1 mg/L (Brain et al. 2004b; Halling-Sørensen 2000; Holten Lützhøft et al. 1999). It is interesting to note that in an early review of antibiotic inhibition of plant growth, penicillin was found to be by far the least toxic (Brian 1957).

Recent research has indicated penicillin-binding proteins (PBPs) in 12 cyanobacterial genomes that are involved in the synthesis of the peptidoglycan layer of the cell wall during cyanobacterial cellular differentiation (Leganés et al. 2005). Comparatively, among plants, glaucophytes (mosses) have retained peptidoglycans, red algae have no genes for peptidoglycan synthesis, and green plants have retained some peptidoglycan genes as indicated by genomic analyses with *A. thaliana*, though their functions have not been determined (Machida et al. 2006). No wall-like structures have been detected in the plastids of green plants, and knockouts of these genes showed no effect on chloroplast division (Machida et al. 2006). Cephalexin, a carbacephem antibiotic with a similar mode of action, also showed no significant effect on *L. gibba* below 1 mg/L (Brain et al. 2004b), although no data are available at present for algae. Monensin is an ionophore antibiotic that affects bacteria mitochondria by disrupting cell membranes and allowing excess water into organelles (Sweetman 2002). Although no toxicity data are available for algae, monensin is moderately toxic to *L. gibba*, with an EC_{50} of 998 µg/L (Brain et al. 2004b).

F. Other Classes of Pharmaceuticals: Mevalonic Acid Pathway Inhibitors

Statin class pharmaceuticals (e.g., lovastatin and atorvastain) are inhibitors of 3-hydroxy-3-methylglutaryl coenzyme-A reductase, the rate-determining step in the cholesterol biosynthetic pathway of mammals (Sweetman 2002) and the mevalonic acid (MVA) pathway in higher plants, which, under normal physiological conditions, is responsible for the formation of sterols and ubiquinone (Disch et al. 1998a; Lichtenthaler et al. 1997a). Statins have demonstrated phytoxicty and sterol reductions in *L. gibba* as far back as 1982 (Bach and Lichtenthaler 1982b); such sterol responses have more recently been observed in a concentration-dependent manner (Brain et al. 2006a). Statin concentrations as low as 36 µg/L are sufficient to reduce sterol concentrations by 50% in *L. gibba*, inhibiting growth of new fronds (Brain et al. 2006a). Statins inhibit sterol formation in radish plants by up to 84% at 250 µM (101 mg/L) (Bach and Lichtenthaler 1983). A concentration of 5 µM (2 mg/L) of lovastatin is sufficient to completely shut down the MVA pathway in tobacco BY-2 cells, and 0.1 µM (40 µg/L) reduces fresh weight by >50% (Disch et al. 1998a; Hemmerlin et al. 2003).

In plants, sterols reinforce membranes regulating acyl chain ordering and water permeability of the phospholipid bilayer, where alterations in plasma membrane sterol profiles can affect or modulate functions of membrane-bound proteins such as enzymes, channels, receptors, or components of signal transduction pathways (Grandmougin-Ferjani et al. 1997; Schaller 2004; Shuler et al. 1991). Sterols are also crucial to cellulose synthesis in the building of the plant cell wall and necessary for cellular division and expansion, embryogenesis, and normal plant growth and development (Jang et al. 2000; Schrick et al. 2000). However, green algae such as *S. obliquus* are insensitive to statins because of the absence of the target pathway (Schwender et al. 1996). In higher plants, there are two independent pathways for synthesizing isopentyl diphosphate (IPP) and its isomer dimethylallyl phosphate (DMAPP), the universal precursors to isoprenoid biosynthesis. However, the statin-sensitive cytosolic MVA pathway of eukaryotic origin is absent in green algae, which synthesize all isoprenoids exclusively from the chloroplastic (2-C-methyl-D-erythritol 4-phosphate, MEP) pathway of prokaryotic origin (Disch et al. 1998a; Lichtenthaler et al. 1997b; Schwender et al. 1996).

Although plants contain both pathways, and some metabolic cross-talk does occur largely through IPP and, to a lesser extent, DMAPP, geranylgeranyl diphosphate, and farnesyl diphosphate (Bick and Lange 2003; Hemmerlin et al. 2003; Laule et al. 2003), these pathways have a finite potential to compensate each other before toxic effects are realized (Brain et al. 2006a; Hemmerlin et al. 2003). No literature sources were found detailing the isoprenoid biosynthetic pathway in blue-green algae; however, in malaria parasites such as *P. falciparum* (of the phylum *Apicomplexa*), isoprenoids are also synthesized via the MVA-independent MEP pathway and shown to be insensitive to lovastatin at pharmacologically relevant concentrations (Wiesner et al. 2004).

G. Other Classes of Pharmaceuticals: General

Generally, apart from antibiotics and statin-type blood lipid regulators, there are few other therapeutic classes of pharmaceuticals that exert significant toxicity to plants. Therefore, beyond a few notable exceptions, these classes likely have effects via narcosis and are not be elaborated on here. One exception includes the selective serotonin reuptake inhibitors (SSRIs). Among the SSRI class, there are varying degrees of potency toward green algae; where EC_{50}s for fluoxetine and sertraline range from 24 to 4339.25 μg/L and 12.10 to 763.66 μg/L, respectively (Brooks et al. 2003; Johnson et al. 2006). Green algae were less sensitive to fluvoxamine, which showed EC_{50}s above 3,000 μg/L for all species tested (Johnson et al. 2006). For both fluoxetine and sertraline, *S. capricornutum* was the most sensitive and *Chlorella vulgaris* the least sensitive algal species tested, although no blue-green species were included. When tested as a mixture of fluoxetine, sertraline, and fluvoxamine in aquatic microcosms, an algal total abundance EC_{50} of 601.0 nM was obtained for 7 d exposure. As algal populations recovered, this estimate increased to 1057.8 nM by 35 d exposure (Johnson et al. 2006). Johnson et al. (2006) suggest a possible mode-specific response in algae because toxicity estimates for SSRIs based on ecological structure–activity relationship models were greater than experimental values, although no biological premise was elaborated. Higher plants appear less sensitive to both fluoxetine and sertraline; tests with *L. gibba* indicated EC_{50}s above 1,000 µg/L for both compounds (Brain et al. 2004a,b).

No pharmaceutical belonging to nonsteroidal antiinflammatory drug (NSAID), analgesic, antiepileptic, fibrate-type lipid regulator, β-blocker, vasodilator, antihyperglycaemic, antimetabolite, nitroimidazoles, nicotine, or stimulant therapeutic classes of pharmaceuticals tested to date have shown any significant toxicity toward aquatic plants at exposures below 500 µg/L. These results suggest that for a pharmaceutical compound to pose a significant risk to plants, the presence of an appropriate receptor is a key requirement. However, the mere presence of a receptor does not guarantee toxicity because the pharmaceutical must reach the receptor at a sufficient internal dose, which requires considerations of uptake, translocation, and/or metabolism.

VII. Differential Sensitivity to Pharmaceuticals

A. Comparative Susceptibility of Higher Plants and Blue-Green and Green Algae

One trend readily apparent when comparing Tables 1 and 2 is the striking degree of class-specific differential sensitivity between blue-green algae, green algae, and *Lemna* sp. For example, the higher plant *Lemna* sp. and cyanobacteria such as *M. aeruginosa* demonstrate similar sensitivity to fluoroquinolones; however, green algae such as *S. capricornutum* are considerably less sensitive by as much as a factor of 2. Contrastingly, *Lemna* sp. had comparable sensitivity to both blue-green and green algae with respect to tetracyclines and remarkably similar sensitivity with respect to sulfonamide antibiotics. However, *Lemna* was found to be the least sensitive to macrolide antibiotics, whereas *M. aeruginosa* was the most sensitive, and *S. capricornutum* displayed variable sensitivity within this class, with EC_{50} values spanning two orders of magnitude. Aminoglycosides show similar sensitivity trends as macrolides, in that both *M. aeruginosa* and *S. capricornutum* demonstrate sensitivity to streptomycin, whereas *Lemna* is typically at least an order of magnitude less sensitive. Both green and blue-green algae as well as diatoms are highly sensitive to triclosan, although *Lemna* was less sensitive at tested concentrations.

SSRIs are also potent toxicants to a number of green algal species, although higher plants appear much less sensitive by at least an order of magnitude. Blue-green algae are highly sensitive to β-lactams, although green algae and higher plants are comparatively insensitive by an order of magnitude. Neither algae nor higher plants show a high degree of sensitivity to pharmaceutical belonging to the nonsteroidal antiinflammatory drug (NSAID), analgesic, antiepileptic, fibrate-type lipid regulator, β-blocker, vasodilator, or antihyperglycemic therapeutic classes of pharmaceuticals, which have no known receptors in plants. It is interesting to note, however, that in tests with chlorophytic algae (*Desmodesmus subspicatus*), and *L. minor* exposed to clofibrinic acid (fibrate-type lipid regulator), carbamazepine (an antiepileptic), diclofenac, naproxen, and ibuprofen (NSAIDs), captopril (antihypertensive), metformin (antidiabetic), propranolol, and metoprolol (β-blockers), *Lemna* was found to be the most sensitive test species in the majority of all tested compounds, with EC_{50} values ranging from 7.5 to >320 mg/L (Cleuvers 2003). Because blue-green and green algae as well as higher plants present receptors to a number of pharmaceuticals, particularly antibiotics, it is a reasonable postulate that differences in sensitivity therefore lie in different metabolic potentials of these organisms. This suggestion has also been put forward with respect to the differential sensitivity demonstrated between algal species and higher plants exposed to a variety of herbicides.

Similar arrays in sensitivity can be seen in *L. gibba* and green algae treated with a number of herbicides, where receptors are known in both species, although the susceptibility varies markedly depending on the class (Fairchild et al. 1997; Mohammad et al. 2005). Fairchild et al. (1997) tested the comparative sensitivity of *L. minor* and *S. capricornutum* to 16 different herbicides and found that results were not always predictable in spite of obvious differences in herbicide mode of action and plant phylogeny (Fairchild et al. 1997). Both plant species were generally sensitive to triazines (photosystem II inhibitors), sulfonureas (branched-chain amino acid inhibitors), pyridines (photosystem I inhibitors), dinitroaniline (microtubule assembly inhibitors), and acetanilide (protein synthesis inhibitors), although *Lemna* was more sensitive to the sulfonureas and the pyridines than *S. capricornutum*, and *S. capricornutum* was more sensitive to one of two thiocarbamates (lipid synthesis inhibitor) and one of the triazines (Fairchild et al. 1997).

Similar departures in sensitivity of *S. capricornutum* and *Lemna* also occurred between chemicals within individual classes of the triazine, acetanilide, and thiocarbamate herbicides, which was suggested to be the result of differential abilities of the two species of plants to metabolize individual chemicals within herbicide classes, although the differences among these metabolic capabilities are not fully understood (Fairchild et al. 1997). Comparisons in sensitivity of five species of aquatic higher plants and six species of algae to triazine and acetanilide herbicides indicated that green algae (*Selenastrum*, *Chlorella*, *Chlamydomonas*, and *Scenedesmus*) were typically more sensitive than the blue-green algae (*Microcystis* and *Anabaena*) (Fairchild et al. 1998). *Lemna* was found to have intermediate sensitivity compared to *Ceratophyllum*, *Najas*, *Elodea*, and *Myriophyllum* and similar sensitivity compared to *Selenastrum* (Fairchild et al. 1998). Other studies have also shown similar trends. For example, treatment of eight sulfonylurea herbicides to both *Lemna* sp. and *P. subcapitata* (*S. capricornutum*) showed sensitivity ranging from 0.91 to 14.5 and 0.27 to >1,000 μg/L, respectively, where *Lemna spp.* was more sensitive to five compounds, *P. subcapitata* was more sensitive to one, and two compounds indicated similar sensitivity (Mohammad et al. 2005). Again, differences in sensitivity were suggested to result from different metabolic abilities of the species (Mohammad et al. 2005).

Many herbicides exert selectivity between different plant species by exploiting the different metabolic potentials of "weed" and "crop" species; for example, maize tolerance to triazine and chloroacetamide herbicides is attributed to rapid detoxification by glutathione transferases (BPCP 2006). Metabolism in plants typically involves oxidation, reduction, and hydrolysis reactions by enzymes such as cytochrome P-450s (phase I) followed by conjugation and secondary conjugation to glucose, amino acids, and glutathione (phase I and II) (Shimabukuro 1985). It is also interesting to note that the range of EC_{50} values obtained for investigations of the effects of herbicides on both higher plants and algae are highly similar to the range of pharmaceutical potencies demonstrated for the same species, suggesting similar arrays in metabolic potential and a herbicidal potential for a number of these compounds.

B. Enzymes Involved in Metabolizing Xenobiotics: Differences Among Species

Plants are equipped with a remarkably versatile system that protects them from the potentially phytotoxic actions of xenobiotics, although there is extraordinary diversity among species, where certain plants have a natural tolerance toward compounds that profoundly affect closely related species (Kreuz et al. 1996). Plants contain a multitude of enzymes that metabolize drugs, herbicides, and other xenobiotics to nonphytotoxic products, primarily a family of enzymes known as cytochrome P-450s or monooxygenases, which are microsomal heme-thiolate proteins similar to those characterized in the endoplasmic reticulum of mammalian livers (Kreuz et al. 1996; Paquette et al. 2000). P-450s typically catalyze the primary step in the detoxification process, a metabolic attack through oxidation or hydrolysis, which usually serves to introduce a hydroxy, amino, or carboxylic acid functional group that is suitable for subsequent conjugation to an endogenous moiety; for example, sugars or thiols (GSH) facilitated by glucosyltransferases and glutathione S-tranferases (GSTs) (Kreuz et al. 1996). As described by Kreuz et al. (1996), the resulting conjugates are (a) generally inactive toward the initial target site; (b) more hydrophilic and less mobile in the plant than the parent herbicide; and (c) susceptible to further processing, which may include secondary conjugation, degradation, and compartmentation.

Cytochrome P-450s are present in all kingdoms, although in bacteria, they are soluble, whereas in eukaryotes they are membrane anchored (Paquette et al. 2000). In addition to their role in the detoxification process, plant P-450s also participate in a myriad of biochemical pathways including those devoted to the synthesis of plant products, such as phenylpropanoids, alkaloids, terpenoids, lipids, cyanogenic glycosides, and glucosinolates, and plant growth regulators, such as gibberellins, jasmonic acid, and brassinosteroids (The Arabidopsis Genome Initiative 2000). However, the role of these endogenous enzymes in plants is generally poorly understood, and only a few dozen P-450 enzymes have been characterized to any extent (The Arabidopsis Genome Initiative 2000). In the superfamily of divergent genes that encode these proteins, 246 full-length, putatively functional P-450 coding sequences represent approximately 1% of the *Arabidopsis* gene complement compared to a range of between 54 and 105 P-450s in the human, mouse, *Takifugu rubripes*, *Anopheles gambiae*, *Drosophila melanogaster*, *Caenorhabditis elegans*, *Ciona intestinalis*, and *Ciona savignyigenomes* genomes, representing approximately 0.1%–0.5% of each of these gene complements (Nelson et al. 2004a,b). As of 2004, 1,098 plant P-450s have been named, and projections suggest 10,500 P-450 sequences are needed to completely characterize the compliment of P-450 proteins in higher plants (Nelson et al. 2004a), indicating incredible diversity among the plant kingdom. Current research suggests that there are 10 clans (recognizable clusters of related P-450 families) in plants, where dicots contain 9 families and monocots contain 10; 4 families have currently been identified in *C. reinhardtii* (Nelson et al. 2004a). The redundancy of P-450 families among plants suggests that species-specific variations account for the differential potential for metabolizing xenobiotics.

Similar to P-450s, differences in the properties of GST isoenzymes also account for conspicuous species-specific differences in metabolism and hence for the vast differences in the susceptibility of different plant species toward certain herbicides (Kreuz et al. 1996), and, likely in the context of this review, pharmaceuticals. Striking similarities can be found between plants and animals in their detoxification mechanisms with respect to the involvement of P-450s and conjugation of electrophilic substrates to GSH, although differences are observed for the other major conjugation step in that xenobiotics are often glucuronated in animals but glucosylated in plants (Kreuz et al. 1996).

VIII. Microcosm Investigations on Pharmaceutical Effects

A small number of microcosm studies have been conducted on pharmaceuticals both singularly and in combination. An investigation of a ternary mixture of ibuprofen (antiinflammatory), fluoxetine (selective serotonin reuptake inhibitor), and ciprofloxacin (antibiotic) indicated severe phytotoxicity to the aquatic macrophytes *L. gibba*, *M. sibiricum*, and *Myriophyllum spicatum* at high (600, 1,000, and 1,000 μg/L) and growth inhibition in *L. gibba* at medium (6, 10, and 10 μg/L) concentrations (Richards et al. 2003) when evaluated in 12,000-L microcosms. In the same systems, a subsequent eight-compound mixture (atorvastatin, acetaminophen, caffeine, sulfamethoxazole, carbamazepine, levofloxacin, sertraline, and trimethoprim) was conducted at total (summed) molar concentrations of 0, 0.044, 0.608, 2.664, and 24.538 μmol/L ($n = 3$). Phytotoxicity was demonstrated in both rooted and floating macrophytes with molar EC_{50}s ranging from 4.6 to 66.7 μmol/L for 35-d exposures of *M. sibiricum* and 1.15 to 12.4 μmol/L for 7-day exposures with *L. gibba* (Brain et al. 2004a). In subsequent laboratory tests with *L. gibba*, the phytotoxic agents in the mixture were found to be levofloxacin, sulfamethoxazole, and atorvastatin (Brain et al. 2004a).

Indirect toxicity has been observed for a mixture of tetracycline antibiotics (tetracycline, chlortetracycline, oxytetracycline, and doxycycline) conducted in these facilities as well, where *L. gibba* was unaffected but *M. sibiricum* experienced significant concentration-dependent decreases in a number of endpoints, most likely mediated via reduction in light at the depths where these plants are found (Brain et al. 2005b). In the same study, phytoplankton abundance showed an initial decrease in the 0.218, 0.668, and 2.289 μmol/L treatments after 7 d; however, all treatments rebounded to levels above the controls after 35 d (Wilson et al. 2001). Also conducted at these facilities, a mixture of three SSRIs, fluoxetine, fluvoxamine, and sertraline were found to reduce total algal abundance and number of taxa by 50% at 601 and 1058 nM at 7 d (Johnson et al. 2006). However, after 35 d no differences were found for number of taxa, and the EC_{50} for algal abundance increased to 1057 nM (Johnson et al. 2006).

IX. Pathway- and Receptor-Specific Endpoints in Risk Assessment

Pathway- or receptor-specific endpoints can be considered measurable stressor-induced variations in biochemical, physiological, and morphological status of an organism, which is the working definition of a "biomarker" (Ernst and Peterson 1994; NRC 1987). In a risk assessment framework, it is useful to classify the various sublethal responses plants may exhibit following exposure to pharmaceuticals as biomarkers of exposure or biomarkers of effect (Stegeman et al. 1992). Biomarkers of exposure differ from biomarkers of effect in that these measures inform whether an organism has been exposed to contaminant (e.g., gene expression) but do not necessarily allow for determination of whether the organism has been adversely affected. Biomarkers of effect (e.g., histopathology), however, are indicators of toxicity, especially when measures of an adverse effect can be linked to a physiologically and ecologically relevant endpoint (e.g., growth). Biomarkers of exposure and effect have been successfully employed in a variety of retrospective ecological risk assessments (Suter et al. 2000); however, biomarker use in prospective risk assessment frameworks for pharmaceuticals is lacking.

Measurable biological responses can range from effects at the molecular level, through to effects on the intact organism, to population, community structure, and, potentially, the structure and function of ecosystems (Peakall

1994). As one moves up this continuum, however, it becomes increasingly difficult to relate causation of effects (Peakall 1994). Currently, in ecotoxicology of plants, the endpoints typically employed for risk assessment are largely gross acute endpoints such as growth rate and biomass measurements (e.g., fresh weight) as well as visual symptoms such as chlorosis and necrosis (Davy et al. 2001). However, as Davy et al. (2001) point out, in plants toxicity is first manifested at the biochemical level before effects are evident at the whole-organism level. These effects are typically more sensitive to environmental stressors, although their relevance and relationship with gross parameters such as biomass are not always evident (Davy et al. 2001). Ernst and Peterson (1994) suggest that biomarkers should be selected from the events of biochemical or physiological pathways, and the more specific the effect of an environmental stress on a metabolic process, the greater its diagnostic potential.

There are currently a number of biomarkers of exposure and effect that can indicate the type or even the nature of stress, including phase I and phase II detoxification enzymes (e.g., cytochrome P-450 oxygenases and glutathione S-transferase multifunctional enzymes, respectively) (Pflugmacher et al. 2000; Roy et al. 1995), oxidative stress including production of reactive oxygen species (ROS) (Akhtar et al. 2005; Babu et al. 2003; Xie et al. 2006), increased activity and levels of peroxidase and its isozymes as well as ascorbic acid and gluatathione (Byl and Klaine 1991; Lytle and Lytle 2001), chlorophyll fluorescence (Marwood et al. 2001), flavonoid production (Wilson et al. 2001), production of phenolic compounds (Ferrat et al. 2003), and heat shock proteins (Akhtar et al. 2005; Lewis et al. 1998, 2001). However, pharmaceuticals are highly specific biologically active compounds designed to effect or influence a particular biological process in a particular target organism or group of organisms. Therefore, it is more appropriate to assess the effects, and subsequently the risk, of these compounds to plants by evaluating specific pathway or target endpoints, provided the pathway or receptor is present.

This approach requires an in-depth understanding of the mode of action in the target organism and the evolutionary context and premise for the presence of the associated pathway or receptor in the nontarget organism. An example of this has recently been demonstrated with statin-type blood lipid regulators in *L. gibba*, where measuring metabolites downstream (sterols) of the target enzyme (HMG-CoA reductase) provided a specific "biomarker" that showed sensitivity two to three times lower than fresh weight (Brain et al. 2006a). Another example of increased sensitivity and specificity of pathway- or- receptor-specific responses has been demonstrated with the herbicide isoxyflutole, a carotenoid biosynthetic inhibitor, where treatment of cucumber cotyledons with 5 mM isoxaflutole reduced metabolite (carotenoid) levels downstream of the target enzyme (*p*-hydroxyphenyl pyruvate dioxygenase) by nearly twice the proportion of fresh weight reduction (Kushwaha and Bhowmik 1999). Thus, for pathway- or receptor-specific biomarkers to be useful, there must be a plausible biochemical, biological, and physiological premise for the effects, and the pathway- or receptor-specific response should directly relate to effects higher up on the continuum to the intact organism. In the case of statin-induced reduction of sterol concentrations, sterols perform a multitude of critical functions in plants, providing a logical relatedness of effect to the morphological endpoints. Similarly, for isoxyflutole, carotenoids are crucial for photosynthesis in plants, and reductions of this metabolite caused predictable ramifications at the morphological level.

In the context of pharmaceuticals and plants, a target- or pathway-specific biomarker should be reflected by effects at the morphological level and accompanied by a compelling biochemical rationale directly relating the two. Furthermore, highly specific biomarkers will only serve a greater utility if they increase the diagnostic resolution, provide mechanistic-based data, provide greater sensitivity than an associated morphological effect, are more specific to the stressor of interest, and, most importantly, if they are relevant. No single biochemical, physiological, or morphological endpoint is in and of itself practical and/or useful because effects must be related and assessed across a continuum of organization. These "alternative" testing endpoints are not considered in regulatory data submissions for pesticides, and it is considered an area that requires guidance and/or further research efforts before these endpoints can be utilized in aquatic assessment methods (Davy et al. 2001). However, in terms of the risk assessment of both pharmaceuticals and pesticides, pathway- or target-specific endpoints should be considered for incorporation into both prospective and retrospective risk assessment approaches, provided the foregoing considerations are established.

X. Risk

All toxicity data for pharmaceuticals in aquatic plants reported to date generally indicate that the risk of adverse effects from exposures at environmentally relevant concentrations is generally low, although there are exceptions. Currently, risk assessments for pharmaceuticals typically rely on hazard or risk quotient (HQ or RQ) approaches, where a predicted environmental concentration (PEC) or measured environmental concentration (MEC) is divided by a statistically derived toxicological benchmark concentration (NOEC or ECx value), usually under worst case circumstances. HQ or RQ ratios greater than 1 indicate potential hazard or "risk," and values less than 1 indicate low potential hazard or risk. Under the USFDA guidance document for the environmental assessment of pharmaceuticals (FDA 1998), a PEC value of 0.10 μg/L is suggested (1.0 μg/L effluent concentration threshold times 10-fold standard dilution factor), and tier 2 assessments (acute ecotoxicity testing on a base set of aquatic organisms) suggest dividing the EC_{50} for the most sensitive organism by a factor of 100, providing a HQ equation of 0.1 μg/L/EC_{50}/100. When this equation is applied to the data in Tables 1 and 2, there are no instances of significant hazard or risk for aquatic higher plants, although 12% of the toxicity values indicate a potential for hazard or risk for algae, largely blue-green algae exposed to antibiotics. Figures 2 and 3 show the hazard quotients by class of blue-green algae, green algae, and higher plants exposed to antibiotics and other pharmaceuticals, respectively. Blue-green algae are consistently the most sensitive, accounting for 80% of the HQ exceedence incidences, where *M. aeruginosa* indicated potential risks from exposure to aminoglycoside, pleuromutilin, macrolide, β-lactam, quinolone, and tetracycline antibiotics and *A. flos-aquae* indicated potential risks from exposure to triclosan. *S. capricornutum* and *S. subspicatus* (species of green algae) also showed potential risks from exposure to triclosan, as well as one macrolide compound (clarithromycin) in the case of *S. capricornutum.*

It is not surprising that blue-green algae display the greatest sensitivity across the spectrum of antibiotics because they are the most bacterial in nature with the most rudimentary barriers for uptake and metabolic capacity. However, the general consensus from the conclusions of the referenced literature, from which the toxicity database was compiled, largely supports the conclusion of low risk, even when higher tier (microcosm) studies (Brain et al. 2004a, 2005a,b), probabilistic approaches and pathway/receptor specific endpoints are employed (Brain et al. 2006a,b). Notwithstanding the general trends, there does appear to be some legitimate concern with respect to certain antibiotics, largely to species of blue-green algae, particularly triclosan, which also poses potential risks to species of green algae (See Fig. 2). The only other instance concerns the indirect effects of tetracycline on light penetration and submerged macrophytes, although assessed under the USFDA HQ paradigm outlined above, the HQ values are below 1.

A number of uncertainties need to be addressed with respect to the risk pharmaceuticals pose to aquatic plants. The toxicity data compiled in this review only encompass slightly more than 2% of the 4,500 active parent pharmaceutical compounds, although the dataset does represent most of the heavily prescribed and discharged pharmaceuticals, providing greater confidence in the risk conclusions. There is also a great deal of uncertainty with respect to sublethal responses and whether pathway- or receptor-mediated responses can be related to physiological effects while providing a more sensitive and specific indication of risk.

Summary

Pharmaceuticals are biologically active, ubiquitous, low-level contaminants that are continuously introduced into the environment from both human and veterinary applications at volumes comparable to total pesticide loadings. Recent analytical advances have made possible the detection of a number of these compounds in environmental samples, raising concerns

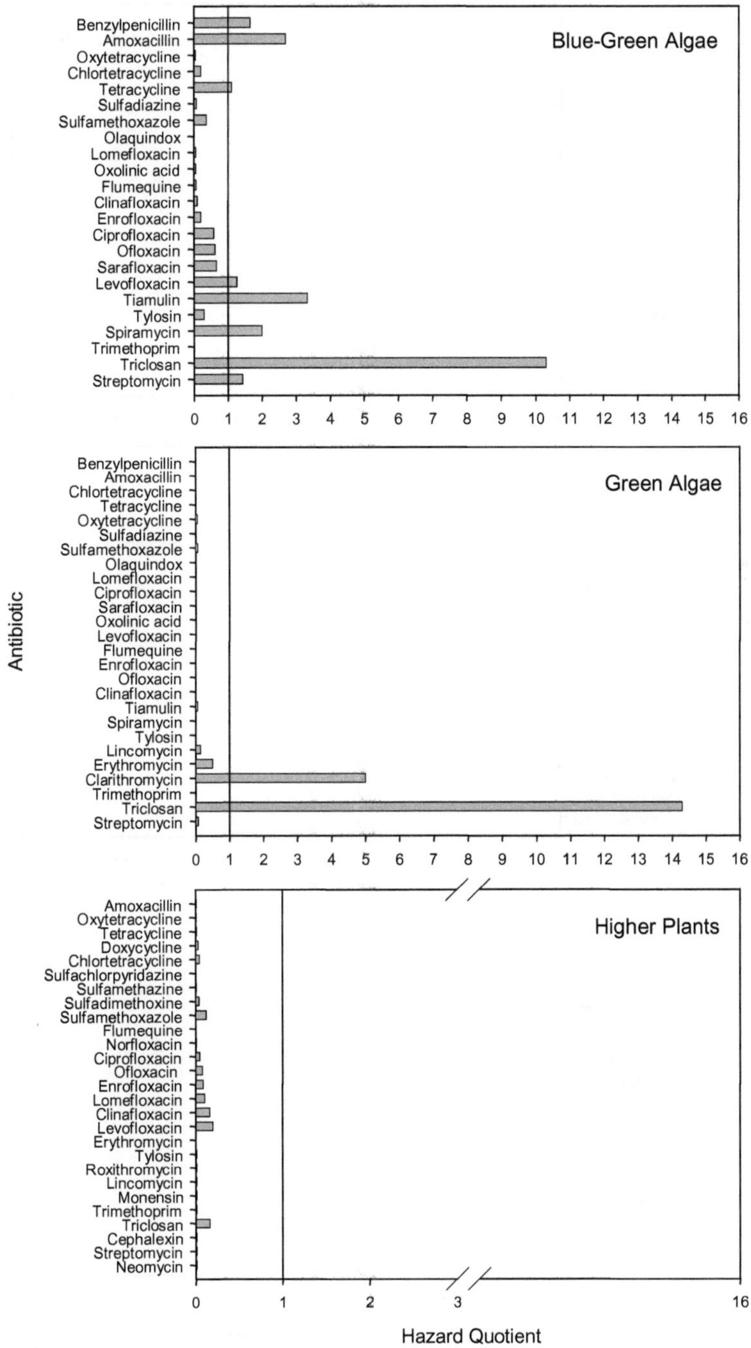

Fig. 2. Hazard quotients (HQs) calculated for blue-green algae, green algae, and higher plants exposed to antibiotics. The HQs were calculated as the ratio of the predicted environmental concentration (PEC: 0.10 μg/L) and the lowest EC50 value or lowest toxicity value available for each compound. The critical HQ of 1 is indicated by a reference line.

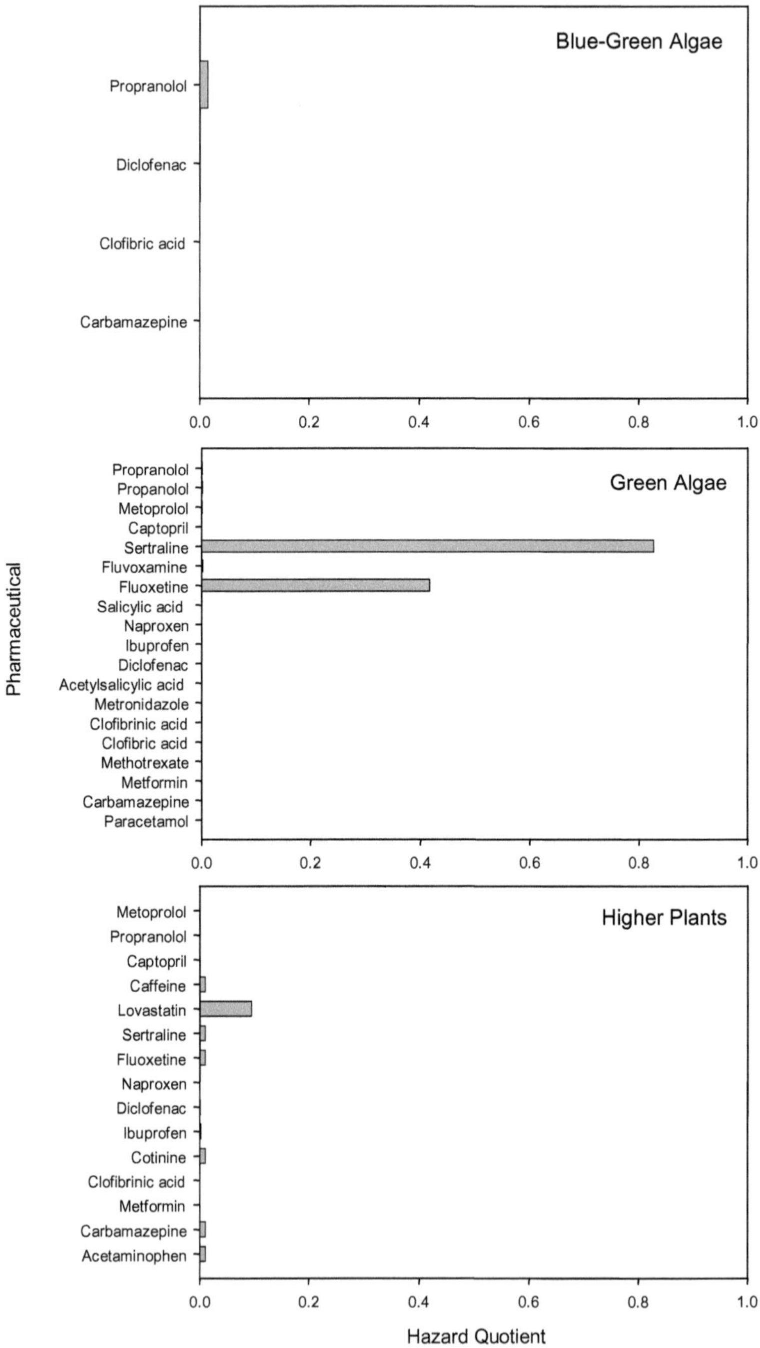

Fig. 3. Hazard quotients (HQs) calculated for blue-green algae, green algae and higher plants exposed to other pharmaceuticals. The HQs were calculated as the ratio of the predicted environmental concentration (PEC: 0.10 μg/L) and the lowest EC50 value or lowest toxicity value available for each compound.

over potential nontarget effects to aquatic organisms, especially given the highly specific biologically active nature of these compounds. These concerns become paramount when the evolutionary conservation of metabolic pathways and receptors is taken into consideration, particularly in the case of aquatic plants, where a great deal of homology is displayed between the chloroplast and bacteria, as well as between other metabolic pathways across multiple phyla of biological organization. Common receptors have been identified in plants for a number of antibiotics affecting chloroplast replication (fluoroquinolones) transcription and translation (tetracyclines macrolides, lincosamides, β-aminoglycosides, and pleuromutilins), metabolic pathways such as folate biosynthesis (sulfonamides) and fatty acid biosynthesis (triclosan), as well as other classes of pharmaceuticals that affect sterol biosynthesis (statin-type blood lipid regulators). Toxicological investigations into the potency of these compounds indicates susceptibility across multiple plant species, although sensitivity to these compounds varies widely between blue-green algae, green algae, and higher plants in a rather inconsistent manner, except that Cyanobacteria are largely the most sensitive to antibiotic compounds. This differential sensitivity is likely dependent on differences in metabolic potential as well as uptake kinetics, which has been demonstrated for a number of compounds from another class of biologically active compounds, pesticides. The demonstration of conserved receptors and pathways in plants is not surprising, although it has been largely overlooked in the risk assessment process to date, which typically relies heavily on physiological and/or morphological endpoints for deriving toxicity data. However, a small number of studies have indicated that measuring the response of a pathway- or receptor-specific target in conjunction with a physiological endpoint with direct relatedness can yield sublethal responses that are two to three times more sensitive that the traditional gross morphological endpoints typically employed in risk assessment. The risk assessment for this review was based almost entirely on evaluations of gross morphological endpoints, which generally indicated that the risk pharmaceuticals pose to aquatic plants is generally low, with a few exceptions, particularly blue-green algae exposed to antibiotics, and both green and blue-green algae exposed to triclosan. It is critical to note, however, that the application of sublethal pathway or receptor-specific responses in risk assessment has largely been unconsidered, and future research is needed to elucidate whether evaluating the toxicity of pharmaceuticals using these endpoints provides a more sensitive, subtle, yet meaningful indication of toxicity than the traditional endpoints used in prospective and retrospective risk assessments for aquatic plants.

Acknowledgments

This work was funded through the Natural Science Engineering Research Council of Canada (NSERC) postdoctoral fellowship program and Baylor

University. We acknowledge contributions from Dr. Paul Sibley, Dr. Jim Bestari, Dr. Linda Lissemore, Dr. Hans Sanderson, Dr. Sean Richards, Christian Wilson, and David Johnson.

References

Adams MS, Titus JE, McCracken MD (1974) Depth distribution of photosynthetic activity in a *Myriophyllum spicatum* community in Lake Wingra. Limnol Oceanogr 19:377–389.

Akhtar TA, Lampi MA, Greenberg BM (2005) Identification of six differentially expressed genes in response to copper exposure in the aquatic plant *Lemna gibba* (duckweed). Environ Toxicol Chem 24:1705–1715.

Alberts AW, Chen J, Kuron G, Hunt V, Huff J, Hoffman C, Rothrock J, Lopez M, Joshua H, Harris E, Patchett A, Monaghan R, Currie S, Stapley E, Albers-Schonberg G, Hensens O, Hirshfield J, Hoohsteen K, Liesch J, Springer J (1980) Mevinolin: a highly potent competitive inhibitor of hydroxymethylglutaryl-coenzyme A reductase and cholesterol-lowering agent. Proc Natl Acad Sci USA 77:3957–3961.

Arigoni D, Sagner S, Latzel C, Eisenreich W, Bacher A, Zenk MH (1997) Terpenoid biosynthesis from 1-deoxy-D-xylulose in higher plants by intramolecular skeletal rearrangement. Proc Natl Acad Sci USA 94:10600–10605.

Babu ST, Akhtar TA, Lampi MA, Tripuranthakam S, Dixon GD, Greenberg BM (2003) Similar stress responses are elicited by copper and ultraviolet radiation in the aquatic plant *Lemna gibba*: implication of reactive oxygen species as common signals. Plant Cell Physiol 44:1320–1329.

Bach TJ, Lichtenthaler HK (1982a) Mevinolin: a highly specific inhibitor of microsomal 3-hydroxy-3-methylglutaryl-coenzyme A reductase of radish plants. Z Naturforsch C 37:46–50.

Bach TJ, Lichtenthaler HK (1982b) Inhibition of mevalonate biosynthesis and of plant growth by the fungal metabolite mevinolin In: Wintermanns JFGM, Kuiper PJC (eds) Biochemistry and Metabolism of Plant Lipids. Elsevier, Amsterdam, pp 515–521.

Bach TJ, Lichtenthaler HK (1983) Inhibition by mevinolin of plant growth, sterol formation and pigment accumulation. Physiol Plant 59:50–60.

Basset GJC, Quinlivam EP, Gregory JF, Hanson AD (2005) Folate synthesis and metabolism in plants and prospects for biofortification. Crop Sci 45:449–453.

Bick JA, Lange BM (2003) Metabolic cross talk between cytosolic and plasidial pathways of isoprenoid biosynthesis: unidirectional transport of intermediates across the chloroplast envelope membrane. Arch Biochem Biophys 415:146–154.

Bock R (2001) Transgenic plastids in basic research and plant biotechnology. J Mol Biol 312:425–438.

Boyd G, Reemtsam H, Grimm D, Mitra S (2003) Pharmaceuticals and personal care products (PPCPs) in surface and treated waters of Louisiana, USA and Ontario, Canada. Sci Total Environ 311:135–149.

BPCP (2006) The e-Pesticide Manual, vol. 2.1. British Crop Protection Agency. Accessed 08/06.

Brain RA, Johnson DJ, Richards RA, Hanson ML, Sanderson H, Lam MW, Young C, Mabury SA, Sibley PK, Solomon KR (2004a) Microcosm evaluation of the effects of an eight pharmaceutical mixture to the aquatic macrophytes *Lemna gibba* and *Myriophyllum sibiricum*. Aquat Toxicol 70:23–40.

Brain RA, Johnson DJ, Richards SM, Sanderson H, Sibley PK, Solomon KR (2004b) Effects of 25 pharmaceutical compounds to *Lemna gibba* using a seven-day static renewal test. Environ Toxicol Chem 23:371–382.

Brain RA, Bestari KT, Sanderson H, Hanson ML, Wilson CJ, Johnson DJ, Sibley PK, Solomon KR (2005a) Aquatic microcosm assessment of the effects of tylosin on *Lemna gibba* and *Myriophyllum spicatum*. Environ Pollut 133: 389–401.

Brain RA, Wilson CJ, Johnson DJ, Sanderson H, Bestari B-J, Hanson ML, Sibley PK, Solomon KR (2005b) Toxicity of a mixture of tetracyclines to *Lemna gibba* and *Myriophyllum sibiricum* evaluated in aquatic microcosms. Environ Pollut 138:426–443.

Brain RA, Reitsma TS, Lissemore LI, Bestari B-J, Sibley PK, Solomon KR (2006a) Herbicidal effects of statin pharmaceuticals in *Lemna gibba*. Environmental Science & Technology 40:5116–5123.

Brain RA, Sanderson H, Sibley PK, Solomon KR (2006b) Probabilistic ecological hazard assessment: evaluating pharmaceutical effects on aquatic higher plants as an example. Ecotoxicol Environ Saf 64(2):128–135.

Brian PW (1957) Effects of antibiotics on plants. Annu Rev Plant Physiol 8:413–426.

Brooks B, Riley T, Taylor R (2006) Water quality of effluent-dominated stream ecosystems: ecotoxicological, hydrological, and management considerations. Hydrobiologia 556:365–379.

Brooks BW, Foran CM, Richards SM, Weston J, Turner PK, Stanley JK, Solomon KR, Slattery M, La Point TW (2003) Aquatic ecotoxicology of fluoxetine. Toxicol Lett 142:169–183.

Brooks DR, Wang P, Read M, Watkins WM, Sims PFG, Hyde JE (1994) Sequence variation of the hydroxymethyldihydropterin pyrophosphkinase; dihydropteroate synthase gene in lines of the human malaria parasite, *Plasmodium falciparum*, with differing resistance to sulfadoxine. Eur J Biochem 224:397–405.

Byl TD, Klaine SJ (1991) Peroxidase activity as an indicator of sublethal stress in the aquatic plant *Hydrilla verticillata* (Royle) In: Gorsuch JW, Lower WR, Wang W, Lewis MA (eds) Plants for Toxicity Assessment, vol. 2. ASTM STP 1115. Amercian Society for Testing and Materials, Philadelphia, pp 101–106.

Champoux JJ (2001) DNA topoisomerases: structure, function, and mechanism. Annu Rev Biochem 70:369–413.

Cleuvers M (2003) Aquatic ecotoxicity of pharmaceuticals including the assessment of combination effects. Toxicol Lett 142:185–194.

Cleuvers M (2004) Mixture toxicity of the anti-inflammatory drugs diclofenac, ibuprofen, naproxen, and acetylsalicylic acid. Ecotox Environ Safe 59:309–315.

Daughton CG (2002) Environmental stewardship and drugs as pollutants. Lancet 360:1035–1036.

Daughton CG, Ternes TA (1999) Pharmaceuticals and personal care products in the environment: agents of subtle change? Environ Health Perspect 107:907–938.

Davy M, Petrie R, Smrchek J, Kuchnicki T, Francois D (2001) Proposal to Update Non-target Plant Toxicity Testing Under NAFTA, USEPA, Washington, DC. http://www.epa.gov/scipoly/sap/.

Disch A, Hemmerlin A, Bach TJ, Rohmer M (1998a) Mevalonate-derived isopentyl diphosphate is the biosynthetic precursor of ubiquinone prenyl side chain in tobacco BY–2 cells. Biochem J 331:615–621.

Disch A, Schwender J, Müller C, Lichtenthaler HK, Rohmer M (1998b) Distribution of the mevalonate and glyceraldehyde phosphate/pyruvate pathways for isoprenoid biosynthesis in unicellular algae and the cyanobacterium Synechocystis PCC 6714. Biochem J 333:381–388.

Divo AA, Sartorelly AC, Patton CL, Frank JB (1988) Activity of fluoroquinolone antibiotics against *Plasmodium falciparum in vitro*. Antibact Agents Chemother 32:1182–1186.

Ebringer L (1972) Are plastids derived from prokaryotic micro-organisms? Action of antibiotics on chloroplasts of *Euglena gracilis*. J Gen Microbiol 71:35–52.

Ernst WHO, Peterson PJ (1994) The role of biomarkers in environmental assessment (4). Terrestrial plants. Ecotoxicology 3:180–192.

Fairchild JF, Ruessler DS, Haverland PS, Carlson AR (1997) Comparative sensitivity of *Selenastrum capricornutum* and *Lemna minor* to sixteen herbicides. Arch Environ Contam Toxicol 32:353–357.

Fairchild JF, Ruessler DS, Carlson AR (1998) Comparative sensitivity of five species of macrophytes and six species of algae to atrazine, metribuzin, alachlor, and metolachlor. Environ Toxicol Chem 17:1830–1834.

Falkowski PG, Raven JA (1997) Aquatic Photosynthesis. Blackwell, Oxford.

FDA (1998) Guidance for industry. Environmental assessment of human drug and biologics applications. U.S. Department of Health and Himan Services, Food and Drug Administration, Center for Drug Evaluation and Research (CDER), Center for Biologics Evaluation and Research (CBER), Rockville, MD.

FEDESA (1997) Antibiotics and animals. FEDESA/FEFANA Press release. European Federation of Animal Health, Brussels, Belgium.

Ferrari B, Paxéus N, Giudice RL, Pollio A, Garric J (2003) Ecotoxicological impact of pharmaceuticals found in treated wastewaters: study of carbamazepine, clofibric acid and diclofenac. Ecotox Environ Safe 55:359–370.

Ferrari B, Mons R, Vollat B, Fraysse B, Paxéus N, Giudice RL, Pollio A, Garric J (2004) Environmental risk assessment of six human pharmaceuticals: Are the current environmental risk assessment procedures sufficient for the protection of the aquatic environment? Environ Toxicol Chem 23:1344–1354.

Ferrat L, Pergent-Martini C, Roméo M (2003) Assessment of the use of biomarkers in aquatic plants for the evaluation of environmental quality: application to seagrasses. Aquat Toxicol 65:187–204.

Fleming A (1929) On the antibacterial action of cultures of a *penicillium*, with special reference to their use in the isolation of *B. influenzæ*. Br J Exp Pathol 10:226–236.

Furukawa H, Tsay J-T, Jackowski S, Takamura Y, Rock CO (1993) Thiolactomycin resistance in *Escherichia coli* is associated with the multidrug resistance efflux pump encoded by *emrAB*. J Bacteriol 175:3723–3729.

Garrison, AW, Pope JD, Allen FR (1976) GC/MS Analysis of organic compounds in domestic waste waters. In Keith CH (ed) Identification and analysis of organic pollutants in water. Ann Arbor Science, Ann Arbor, MI pp 517–566.

Giovannoni SJ, Turner S, Olsen GJ, Barns S, Lane DJ, Pace NR (1988) Evolutionary relationships among cyanobacteria and green chloroplasts. J Bacteriol 170: 3584–3592.

Grandmougin-Ferjani A, Schuller-Muller I, Hartmann MA (1997) Sterol modulation of the plasma membrane H+-ATPase activity from corn roots reconstituted into soyabean lipids. Plant Physiol 113:163–174.

Gray MW, Lang BF (1998) Transcription in chloroplasts and mitochondria: a tale of two polymerases. Trends Microbiol 6:1–3.

Greenberg BM, Huang X-D, Dixon DG (1992) Applications of the aquatic higher plant *Lemna gibba* for ecotoxicological risk assessment. J Aquat Ecosys Health 1:147–155.

Halling-Sørensen B (2000) Algal toxicity of antibacterial agents used in intensive farming. Chemosphere 40:731–739.

Halling-Sørensen B, Jørgensen SE (2000) Drugs in the environment. Chemosphere 40:691–699.

Halling-Sørensen B, Nors Nielsen S, Lanzky PF, Ingerslev F, Holten Lutzhoft HC, Jørgensen SE (1998) Occurrence, fate and effects of pharmaceutical substances in the environment: a review. Chemosphere 36:357–393.

Hanson, AD, Roje, S (2001) One-carbon metabolism in higher plants. Annu Rev Plant Physiol 52:119–137.

Hao C, Lissemore L, Nguyen B, Kleywegt S, Yang P, Solomon K (2006) Determination of pharmaceuticals in environmental waters by liquid chromatography/electrospray ionization tandem mass spectrometry. Anal Bioanal Chem 384(2): 505–513.

Harris EH, Burkhart BD, Gillham NW, Boynton JE (1989) Antibiotic resistance mutations in the chloroplast 16s and 23s rRNA genes of *Chlamydomonas reinhardtii*: correlation of genetic and physical maps of the chloroplast genome. Genetics 123:281–292.

Harwood JB (1996) Recent advances in the biosynthesis of plant fatty-acids. Biochim Biophys Acta 1301:7–56.

Hashimoto H, Murakami S (1982) Chloroplast replication and loss of chloroplast DNA induced by nalidixic acid in *Euglena gracilis*. Cell Struct Funct 7:111–120.

Heath RJ, Rubin JR, Holland DR, Zhang E, Snow ME, Rock CO (1999) Mechanism of triclosan inhibition of bacterial fatty acid synthesis. J Biol Chem 274: 11110–11114.

Heberer T (2002) Occurrence, fate, and removal of pharmaceutical residues in the aquatic environment: a review of recent research data. Toxicol Lett 131:5–17.

Hemmerlin A, Hoeffler J-F, Meyer O, Tritsch D, Kagan IA, Grosdemange-Billiard C, Rohmer M, Bach TJ (2003) Cross-talk between the cytosolic mevalonate and the plastidial methylerythritol phosphate pathways in tobacco bright yellow-2 cells. J Biol Chem 278:26666–26676.

Henschel K-P, Wenzel A, Diedrich M, Fliedner A (1997) Environmental hazard assessment of pharmaceuticals. Regulatory Pharmacology and Toxicology 25:220–225.

Hirsch R, Ternes T, Herberer K, Kratz K-L (1999) Occurrence of antibiotics in the aquatic environment. Sci Total Environ 225:109–118.

Högenauer G, Ruf C (1981) Ribosomal binding region for the antibiotic tiamulin: stoichiometry, subunit location, and affinity for various analogs. Antimicrob Agents Chemother 19:260–265.

Holten Lützhøft HC, Halling-Sørensen B, Jørgensen SE (1999) Algal toxicity of antibacterial agents applied in Danish fish farming. Arch Environ Contam Toxicol 36:1–6.

Isidori M, Lavorgna M, Nardelli A, Pascarella L, Parrella A (2005) Toxic and genotoxic evaluation of six antibiotics on non-target organisms. Sci Tot Environ 346:87–98.

Iyer JK, Milhous WK, Cortese JF, Kublin JG, Plowe CV (2001) *Plasmodium falciparum* crossresistance between trimethoprim and pyrimethamine. Lancet 358:1066–1067.

Jang J-C, Fujioka S, Tasaka M, Seto H, Takatsuto S, Ishii A, Aida M, Yoshida S, Sheen J (2000) A critical role of sterols in embryonic patterning and meristem programming revealed by the fackel mutants of *Arabidopsis thaliana*. Genes Dev 14:1485–1497.

Johnson DJ, Sanderson H, Brain RA, Wilson CJ, Solomon KR (2006) Toxicity and hazard of selective serotonin reuptake inhibitor antidepressants fluoxetine, fluvoxamine, and sertraline to algae. Ecotox Environ Safe 67:128–139.

Jones OAH, Voulvoulis N, Lester JN (2001) Human pharmaceuticals in the aquatic environment a review. Environ Technol 22:1383–1394.

Kato M, Nishida I, Kawaguchi A, Yamada M (1987) Effect of thiolactomycin on fatty acid synthesis in higher plants. Plant Cell Physiol 28:851–855.

Khalil I, Rønn AM, Alifrangis M, Gabar HA, Satti GMH, Bygbjerg IC (2003) Dihydrofolate reductase and dihydropteroate synthase genotypes associated with in vitro resistance of *Plasmodium falciparum* to pyrimethamine, trimethoprim, sulfadoxine, and sulfamethoxazole. Am J Trop Med Hyg 68:586–589.

Kolpin DK, Furlong ET, Meyer MT, Thurman ME, Zaugg SD, Barber LB, Buxton HT (2002) Pharmaceuticals, hormones, and other organic wastewater contaminants in U.S. streams, 1999–2000: a national reconnaissance. Environ Sci Technol 36:1202–1211.

Krajcovic J, Ebringer L, Polonyi J (1989) Quinolones and coumarins eliminate chloroplasts from *Euglena gracilis*. Antibact Agents Chemother 33:1883–1889.

Kreuz K, Tommasini R, Martinoia E (1996) Old enzymes for a new job. Plant Physiol 111:349–353.

Kümmerer K (2001) Drugs in the environment: emission of drugs, diagnoistic aids and disinfectants into wastewater by hospitals in relation to other sources: a review. Chemosphere 45:957–969.

Kushwaha S, Bhowmik PC (1999) Inhibition of pigment biosynthesis in cucumber cotyledons by isoxaflutole. Photosynthetica (Prague) 37:553–558.

Lange BM, Rujan T, Martin W, Croteau R (2000) Isoprenoid biosynthesis: the evolution of two ancient and distinct pathways across genomes. Proc Natl Acad Sci USA 97:13172–13177.

Lanzky PF, Halling-Sørensen B (1997) The toxic effect of the antibiotic metronidazole on aquatic organisms. Chemosphere 35:2553–2561.

Laule O, Fürholz A, Chang H-S, Zhu T, Wang X, Heifetz PB, Gruissem W (2003) Crosstalk between cytosolic and plastidial pathways of isoprenoid biosynthesis in *Arabidopsis thaliana*. Proc Natl Acad Sci USA 100:6866–6871.

Leganés F, Blanco-Rivero A, Fernández-Pinas F, Redondo M, Fernández-Valiente E, Fan Q, Lechno-Yossef S, Wolk CP (2005) Wide variation in the cyanobacterial complement of presumptive penicillin-binding proteins. Arch Microbiol 184:234–248.

Levy SB (1998) The challenge of antibiotic resistance. Sci Am 278:46–53.

Lewis S, May S, Donkin ME, Depledge MH (1998) The influence of copper and heatshock on the physiology and cellular stress response of *Enteromorpha intestinalis*. Environ Res 46:421–424.

Lewis S, Donkin ME, Depledge MH (2001) Hsp70 expression in *Enteromorpha intestinalis* (Chlorophyta) exposed to environmental stressors. Aquat Toxicol 51:277–291.

Lichtenthaler HK, Schwender J, Disch A, Rohmer M (1997a) Biosynthesis of isoprenoids in higher plant chloroplasts proceeds via a mevalonate-independent pathway. FEBS (Fed Eur Biochem Soc) Lett 400:271–274.

Lichtenthaler HK, Rohmer M, Schwender J (1997b) Two independent biochemical pathways for isopentyl diphosphate and isoprenoid biosynthesis in higher plants. Physiol Plant 101:643–652.

Lissemore L, Hao C, Yang P, Sibley PK, Mabury S, Solomon KR (2006) An exposure assessment for selected pharmaceuticals within a model watershed in Southern Ontario. Chemosphere 64:717–729.

Lyman H, Jupp AS, Larrinua I (1967) Specific inhibition of chloroplast replication in *Euglena gracilis* by naladixic acid. J Cell Biol 35:726–730.

Lytle JS, Lytle TF (2001) Use of plants for toxicity assessment of estuarine ecosystems. Environ Toxicol Chem 20:68–83.

Machida M, Takechi K, Sato H, Chung SJ, Kuroiwa H, Takio S, Seki M, Shinozaki K, Fujita T, Hasebe M, Takano H (2006) Genes for the peptidoglycan synthesis pathway are essential for chloroplast division in moss. Proc Natl Acad Sci USA 103:6753–6758.

Mackowiak PA (2000) Brief history of antipyretic therapy. Clin Infect Dis 31:154–156.

Maliga P (1998) Two plastid RNA polymerases of higher plants: an evolving story. Trends Plant Sci 3:4–6.

Manuell A, Beligni MV, Yamaguchi K, Mayfield SP (2004) Regulation of chloroplast translation: interactions of RNA elements, RNA-binding proteins and the plastid ribosome. Biochem Soc Trans 23:601–605.

Marwood CA, Solomon KR, Greenberg BM (2001) Chlorophyll fluorescence as a bioindicator of effects on growth in aquatic macrophytes from mixtures of polycyclic aromatic hydrocarbons. Environ Toxicol Chem 20:890–898.

Maxwell A (1997) DNA gyrase as a drug target. Trends Microbiol 5:102–109.

May KR, Druett HA, Packman LP (1969) Streptomycin inhibition of protein synthesis in peas reversed by divalent cations. Nature (Lond) 221:1147–1148.

McFadden GI, Roos DS (1999) Apicomplexan plastids as drug targets. Trends in Microbiology 7:328–333.

McLeod R, Muench SP, Rafferty JB, Kyle DE, Mui EJ, Kirisits MJ, Mack DG, Roberts CW, Samuel BU, Lyons RE, Dorris M, Milhous MK, Rice DW (2001) Triclosan inhibits the growth of *Plasmodium falciparum* and *Toxoplasma gondii* by inhibition of Apicomplexan Fab I. Int J Parasitol 31:109–113.

Metcalfe CD, Koenig BG, Bennie DT, Servos M, Ternes TA, Hirsch R (2003) Occurrence of neutral and acidic drugs in the effluents of canadian sewage treatment plants. Environ Toxicol Chem 22:2872–2880.

Miao X-S, Metcalfe CD (2003) Determination of pharmaceuticals in aqueous samples using positive and negative voltage switching microbore liquid chromatography/electrospray ionization tandem mass spectrometry. J Mass Spectrom 38:27–34.

Miao X-S, Bishay F, Chen M, Metcalfe C (2004) Occurrence of antimicrobials in the final effluents of wastewater treatment plants in Canada. Environ Sci Technol 38:3533–3541.

Migliore L, Cozzolino S, Fiori M (2000) Phytotoxicity to and uptake of flumequine used in intensive aquaculture on the aquatic weed, *Lythrum salicaria* L. Chemosphere 40:741–750.

Mohammad M, Kishimoto T, Itoh K, Suyama K, Yamamoto H (2005) Comparative sensitivity of *Pseudokirchnetiella subcapitata* vs. *Lemna* sp. to eight sulfonylurea herbicides. Bull Environ Contam Toxicol 75:866–872.

Møllegaard NE, Bailly C, Waring MJ, Nielsen PE (2000) Quinoxaline antibiotics enhance peptide nucleic acid binding to double-stranded DNA. Biochemistry 39:9502–9507.

Montandon P-E, Nicolas P, Schürmann P, Stutz E (1985) Streptomycin-resistance of *Euglena gracilis* chloroplasts: identificatio of a point mutation in the 16S rRNA gen in an invariant position. Nucleic Acids Res 13:4299–4310.

Mullet JE (1993) Dynamic regulation of chloroplast transcription. Plant Physiol 103:309–313.

Nelson DR, Schuler MA, Paquette SM, Werck-Reichhart D, Bak S (2004a) Comparative genomics of rice and *Arabidopsis*. Analysis of 727 cytochrome P450 genes and pseudogenes from a monocot and a dicot. Plant Physiol 135:756–772.

Nelson DR, Zeldin DC, Hoffman SMG, Maltais LJ, Wain HM, Nebert DW (2004b) Comparison of cytochrome P450 (CYP) genes from the mouse and human genomes, including nomenclature recommendations for genes, pseudogenes, and alternative-splice variants. Pharmacogenetics 14:1–18.

Neuburger M, Rébeillé F, Jourdain A, Nakamura S, Douce R (1996) Mitochondria are a major site for folate and thymidylate in plants. J Biol Chem 271:9466–9472.

Nickell LG, Gordon PN (1961) Effects of the tetracycline antibiotics, their degradation products, derivatives and some synthetic analogs on the growth of certain plant systems. Antimicrob Agents Ann 1960:588–593.

Nicolas P (1981) Sensitivity of *Euglena gracilis* to chloroplast-inhibiting antibiotics, and properties of antibiotic-resistant mutants. Plant Sci Lett 22:309–316.

NRC (1987) National Research Council Committee on biological markers. Environ Health Perspect 74:3–9.

Orvos DR, Versteeg DJ, Inauen J, Capdevielle M, Rothenstein A, Cunningham V (2002) Aquatic toxicity of triclosan. Environ Toxicol Chem 21:1338–1349.

Paquette SM, Bak S, Feyereisen R (2000) Intron–exon organization and phylogeny in a large superfamily, the paralogous cytochrome P450 genes of *Arabidopsis thaliana*. DNA Cell Biol 19:307–317.

Peakall DB (1994) The role of biomarkers in environmental assessment (1). Introduction. Ecotoxicology 3:157–160.

Pfannschmidt T, Ogrzewalla K, Baginsky S, Sickmann A, Meyer HE, Link G (2000) The multisubunit chloroplast RNA polymerase A from mustard (*Sinapis alba* L.). Integration of a prokaryotic core into a larger complex with organelle-specific functions. Eur J Biochem 267:253–261.

Pflugmacher S, Schwarz S, Pachur HJ, Steinberg CEW (2000) Effects of tributyltin chloride (TBTCl) on detoxication enzymes in aquatic plants. Environ Toxicol 15:225–233.

Pollock K, Barfield DG, Shields R (1983) The toxicity of antibiotics to plant cell cultures. Plant Cell Rep 2:36–39.

Pomati F, Netting AG, Calamari D, Neilan BA (2004) Effects of erythromycin, tetracycline and ibuprofen on the growth of Synechocystis sp. and *Lemna minor*. Aquat Toxicol 67:387–396.

Pro J, Oritz JA, Boleas S, Fernández C, Carbonell G, Tarazona JV (2003) Effect assessment of Antimicrobial pharmaceuticals on the aquatic plant *Lemna minor*. Bull Environ Contam Toxicol 70:290–295.

Rébeillé FD, Macherel D, Mouillon JM, Garin J, Douce R (1997) Folate biosynthesis in higher plants: purification and molecular cloning of a bifunctional 6-hydroxymethyl-7,8-dihydropterinpyrophosphokinase/7,8-dihydropteroate synthase localized in mitochondria. EMBO Rep 16:947–957.

Richards RA, Wilson CJ, Johnson DJ, Castle DM, Lam MW, Mabury SA, Sibley PK, Solomon KR (2003) Effects of pharmaceutical mixtures in aquatic microcosms. Environ Toxicol Chem 23:1035–1042.

Robinson AA, Belden JB, Lydy MJ (2005) Toxicity of fluoroquinolone antibiotics to aquatic organisms. Environ Toxicol Chem 24:423–430.

Rohmer M (1999) Isoprenoids including steroids and carotenoids. In: Cane DE (ed) Comprehensive Natural Product Chemistry, vol. 2. Pergamon Press, Terrytown, pp 45–68.

Roy S, Lindström-Seppä P, Huuskonen S, Hänninen O (1995) Responses of biotransformation and antioxidant enzymes in *Lemna minor* and *Oncorhynchus mykiss* exposed simultaneously to hexachlorobenzene. Chemosphere 30:1489–1498.

Ryall K, Harper JT, Keeling PJ (2003) Plastid-derived type II fatty acid biosynthetic enzymes in chromists. Gene (Amst) 313:139–148.

Samuelsen OB, Lunestad BT, Husevag B, Holleland T, Ervik A (1992) Residues of oxolinic acid in wild fauna following medication in fish farms. Dis Aquat Org 12:111–119.

Sanderson H, Johnson DJ, Reitsma T, Brain RA, Wilson CJ, Solomon KR (2004) Ranking and prioritization of environmental risks of pharmaceuticals in surface waters. Regul Toxicol Phamacol 39:158–183.

Schaller H (2004) New aspects of sterol biosynthesis in growth and development of higher plants. Plant Physiol Biochem 42:465–476.

Schrick K, Mayer U, Horrichs A, Kuhnt C, Bellini C, Dangl J, Schmidt J, Jürgens G (2000) FACKEL is a sterol C-14 reductase required for organized cell division and expansion in *Arabidopsis embryogenesis*. Genes Dev 14:1471–1484.

Schwender J, Seemann M, Lichtenthaler HK, Rohmer M (1996) Biosynthesis of isoprenoids (carotenoids, sterols, prenyl side-chains of chlorophylls and plastoquinone) via a novel pyruvate/glyceraldehyde 3-phosphate non-mevalonate pathway in the green algae *Scenedesmus obliquus*. Biochem J 316:73–80.

Schwender J, Zeidler J, Groner R, Muller C, Focke M, Braun S, Lichtenthaler FW, Lichtenthaler HK (1997) Incorporation of 1-deoxy-D-xylulose into isoprene and phytol by higher plants and algae. FEBS Lett 414:129–134.

Seiler JP (2002) Pharmacodynamic activity of drugs and ecotoxicology: can the two be connected? Toxicol Lett 131:105–115.

Shimabukuro RH (1985) Herbicide physiology in detoxification of herbicides. In: Weed Physiology, vol. 2. CRC Press, Boca Raton, pp 215–240.

Shuler I, Milon A, Nakatani Y, Ourisson G, Albrecht AM, Benveniste P, Hartmann MA (1991) Differential effects of plant sterols on water permiability and on acyl chain ordering of soybean phosphatidylcholine bilayers. Proc Natl Acad Sci USA 88:6926–6930.

Smith S (1994) The animal fatty acid synthase: one gene, one polypeptide, seven enzymes. Fed Am Soc Exp Bio J 8:1248–1259.

Sosiak A, Hebben T (2005) A Preliminary Survey of Pharmaceuticals and Endocrine Disrupting Compounds in Treated Municipal Wastewaters and Receiving Rivers of Alberta. Alberta Environment, Edmonton, Alberta, Canada.

Stegeman JJ, Brower M, DiGiulio RT, Forlin L, Fowler BA, Sanders BM, Van Veld P (1992) Molecular responses to environmental contamination: enzyme and protein as indicators of chemical exposure and effect In: Huggett RJ, Kimerle RA, Mehrle PM, Bergman HL (eds) Biomarkers: Biochemical, Physiological and Histological Markers of Anthropogenic Stress. Lewis, Boca Raton, pp 235–335.

Suter GW II, Efroymson RA, Sample BE, Jones DS (2000) Ecological Risk Assessment for Contaminated Sites. Lewis, Boca Raton.

Swann MM, Blaxter KL, Field HI, Howie JW, Lucas IAM, Millar ELM, Murdoch JC, Parsons JH, White EG (1969) Report of the joint committee on the use of antimicrobials in animal husbandry and veterinary medicine. Her Majesty's Stationery Office, London, UK.

Sweetman SC (ed) (2002) Martindale: The Complete Drug Reference, 33th Ed. Pharmaceutical Press, London.

Tabak HH, Brunch RL (1970) Steroid hormones as water pollutants. I. Metabolism of natural and synthetic ovulation inhibiting hormones by micro-organisms of activated sludge and primary settled sludge. Dev Ind Microbiol 11:367–376.

The Arabidopsis Genome Initiative (2000) Analysis of the genome sequence of the fowering plant *Arabidopsis thaliana*. Nature (Lond) 408:796–815.

Thompson RJ, Mosig G (1985) An ATP-dependent supercoiling topoisomerase of *Chlamydomonas reinhardtii* affects accumulation of specific chloroplast transcripts. Nucleic Acids Res 13:873–891.

U.S. EPA (2006) Technical overview of ecological risk assessment analysis phase: Ecological effects characterization. U.S. Environmental Protection Agency http://www.epa.gov/oppefed1/ecorisk_ders/toera_analysis_eco.htm. Accessed 2006.

van Pel B, Cocito C (1973) Formation of chloroplast ribosomal RNA in *Euglena* incubated with protein inhibitors. Exp Cell Res 78:111–117.

Wall MK, Mitchenall LA, Maxwell A (2004) *Arabidopsis thaliana* DNA gyrase is targeted to chloroplasts and mitochondria. Proc Natl Acad Sci USA 101: 7821–7826.

Waller RF, Ralph SA, Reed MB, Su V, Douglas JD, Minnikin DE, Cowman AF, Besra GS, McFadden GI (2003) A type II pathway for fatty acid biosynthesis presents drug targets in *Plasmodium falciparum*. Antimicrob Agents Chemother 47:297–301.

Wang P, Read M, Sims PFG, Hyde JE (1997) Sulfadoxine resistance in the human malaria parasite *Plasmodium falciparum* is determined by mutations in dihydropteroate synthetase and an additional factor associated with folate utilization. Mol Microbiol 23:979–986.

Wiesner J, Borrmann S, Jomaa H (2004) Fosmidomycin for the treatment of malaria. Parasitol Res 90:S71–S76.

Wilson BA, Smith VH, Denoyelles F, Larive CK (2003) Effects of three pharmaceutical and personal care products on natural freshwater algal assemblages. Environ Sci Technol 37:1713–1719.

Wilson CJ, Brain RA, Sanderson H, Johnson DJ, Bestari KT, Sibley PK, Solomon KR (2004) Structural and functional responses of plankton to a mixture of four tetracyclines in aquatic microcosms. Environ Sci Technol 38:6430–6439.

Wilson KE, Thompson JE, Huner NPA, Greenberg BM (2001) Effects of ultraviolet-A exposure on ultraviolet-B–induced accumulation of specific flavonoids in *Brassica napus*. Photochem Photobiol 73:678–684.

Woelfle MA, Thompson RJ, Mosig G (1993) Roles of novobiocin-sensitive topoisomerases in chloroplast DNA replication in *Chlamydomonas reinhardtii*. Nucleic Acids Res 21:4231–4238.

Woods KM, Nesterenko MV, Upton SJ (1996) Efficacy of 101 antimicrobials and other agents on the development of *Cryptosporidium parvum in vitro*. Ann Trop Med Parasitol 90:603–615.

Xie F, Koziar SA, Lampi MA, Dixon DG, Warren NP, Borgmann U, Huang XD, Greenberg BM (2006) Assessment of the toxicity of mixtures of copper, 9,10-phenanthrenequinone, and phenanthrene to *Daphnia magna*: evidence for a reactive oxygen mechanism. Environ Toxicol Chem 25:613–622.

Yamaguchi K, Subramanian AR (2000) The plastid ribosomal proteins: identification of all the proteins in the 50S subunit of an organelle ribosome (chloroplast). J Biol Chem 275:28466–28482.

Yamaguchi K, Subramanian AR (2003) Proteomic identification of all plastid-specific ribosomal proteins in higher plant chloroplast 30S ribosomal subunit PSRP-2 (U1A-type domains), PSRP-3a/b (ycf65 homologue) and PSRP-4 (Thx homologue). Eur J Biochem 270:190–205.

Yamaguchi K, Beligni MV, Prieto S, Haynes PA, McDonald WH, Yates JR, Mayfield SP (2003) Proteomic characterization of the *Chlamydomonas reinhardtii* chloroplast ribosome: identification of proteins unique to the 70S ribosome. J Biol Chem 278:33774–33785.

Manuscript received November 22; accepted November 27, 2006.

Rev Environ Contam Toxicol 192:117–158

Risk of Waterborne Illness Via Drinking Water in the United States

Kelly A. Reynolds, Kristina D. Mena, and Charles P. Gerba

Contents

Communicated by David M. Whitacre.

Kelly A. Reynolds (✉)
The University of Arizona, Mel and Enid Zuckerman College of Public Health, 1295 N. Martin Ave., Tucson, AZ 85724, USA

K.D. Mena
University of Texas-Houston 1100 N. Stanton, Suite 110, School of Public Health, El Paso, TX 79902, USA

C.P. Gerba
Department of Soil, Water and Environmental Science, University of Arizona, 429 Shantz Bldg #38 Tucson, AZ 84721, USA

I. Introduction

The quality of drinking water in the United States is among the best in the world; however, waterborne disease outbreaks continue to occur, and many more cases of endemic illness are estimated. Documented waterborne disease outbreaks are primarily the result of technological failures or failure to treat the water (Craun et al. 2006). Current federal regulations require that all surface waters used for a drinking water supply be treated to reduce the level of pathogens so as to reduce the risk of infection to 1:10,000 per year (Regli et al. 1991). To achieve this goal, water treatment must, at a minimum, reduce infectious viruses by 99.99% and protozoan parasites by 99.9% (Regli et al. 2003). If *Cryptosporidium* concentrations exceed a certain level in the source water, additional reductions are required. This degree of treatment is usually achieved by a combination of physical processes (coagulation, sedimentation, and filtration) and disinfection (chlorination, ozonation). Filtration is essential for the removal of protozoan parasites due to their resistance to chlorination and ozonation at doses normally used in drinking water treatment (Barbeau et al. 2000; Korich et al. 1990; Rennecker et al. 1999). A variance from filtration is allowed in some cases if the watershed is protected and carefully monitored for protozoan pathogens.

Before finalization of the U.S. Environmental Protection Agency (USEPA) Ground Water Rule in November 2006, disinfection was not required for drinking water from groundwater sources if coliform bacteria were not detected, as long as the source water was not directly under the influence of surface water. The Groundwater Rule, however, requires all municipal groundwater sources to be disinfected, unless they meet certain monitoring and sanitary survey requirements by December 1, 2009 (USEPA 2006a).

Despite the increase in source water treatment requirements in the U.S., with current and newly promulgated regulations, questions remain as to how much illness is caused by microorganisms in drinking water in the U.S. and what additional approaches may be used to further reduce this risk, especially to sensitive subpopulations who may be at greater risk of infection and more serious adverse health outcomes. The objective of this review is to assess current threats to the water supply in the U.S., provide estimates of total drinking water illness, and suggest approaches for risk reduction.

II. Population Impacts of Waterborne Pathogens

A pathogen is a microorganism capable of causing disease in a host. Relative to the microbial population on earth, only a small number are capable of causing disease in humans. Waterborne pathogens are excreted in the feces of humans and transmitted via ingestion. In contrast, water-based pathogens occur naturally in water and are usually not transmitted from person to person (e.g., *Legionella* spp.). This review primarily deals with waterborne pathogens. Currently, more than 140 known microorganisms are recognized as waterborne pathogens.

Waterborne pathogens have emerged in importance for a number of reasons, including (1) an increase in the size of sensitive subpopulations; (2) recognition of the importance of additional health effects, including chronic sequelae; (3) an increase in the importation of foods from developing countries, where poor water quality plays a role in foodborne illness; (4) natural evolution of microbes with increased virulence; and (5) the use of molecular source tracking to improve methods for identification of outbreaks and their sources.

In recent decades there has been a steady growth in the number of sensitive populations, now thought to comprise 20%–25% of the total U.S. population. Sensitive populations include the elderly, the very young, the chronically ill, recipients of immunosuppressive therapies, and pregnant women (Table 1). Studies show that these subpopulations are more likely to be infected and experience increased morbidity and mortality following exposure to microbial pathogens than the general population (Gerba et al. 1996a; Nwachuku and Gerba 2006). Total diarrheal deaths in aged populations (>74 yr) are about 50% compared to less than 5% in those between the ages of 5 and 24 (Lew et al. 1991). Adenovirus infections have proved problematic for immunocompromised populations, in which a 60% and 53% case-fatality rate is prevalent in bone marrow transplant and cancer patients, respectively (Table 2). In addition, *Cryptosporidium* is identified in 2.2% of all diarrhea cases in developed countries compared to a 7% rate in children and 14% (range, of 6%–70%) in acquired immunodeficiency syndrome (AIDS) patients (Chen et al. 2002).

Table 1. Sensitive Subpopulations in the United States.

Number	Population
35,061,000	Persons >65[a]
20,186,469	Children <5[b]
18,200,000	Persons with diabetes[c]
15,000,000	Cancer patients[d]
6,000,000	Pregnant women[e]
1,039,000	AIDS patients[f]
123,120[h]	Organ transplants[g]

[a]Velkoff and DeBarros (2005).
[b]US Census Bureau (2005).
[c]CDC (2005).
[d]Jemal et al. (2005).
[e]American Pregnancy Association (2006).
[f]Glynn and Rhodes (2005).
[g]US Department of Health and Human Services (2005).
[h]Five-year total recipients, 2000–2004.

Table 2. Case-fatality Rates among Immunocompromised Patients with Adenovirus Infection.

Patient group	Overall % fatality	Mean age of patients (yrs)
Bone marrow transplant	60	15.6
Liver transplant	53	2.0
Renal transplant	18	35.6
Cancer patients	53	25.0
AIDS patients	45	31.1

Source: Modified from Hierholzer (1992).

Although diarrhea is the major symptom associated with waterborne pathogens, other chronic sequelae are possible (Parkin 2000). Chronic sequelae are diseases that develop in the days, weeks, or years after initial infection. Chronic sequelae, such as diabetes, heart disease, autoimmune disease, and cancer, can have a significant impact on the individual's quality of life and are sometimes related to infectious disease agents. In addition, exposure to some pathogens can lead to adverse effects on the endocrine system (i.e., *Giardia lamblia* is linked to hypothyroidism and coxsackievirus is linked to orchitis) (Lindsay 1997).

Globalization of commerce and travel contributes to the spread of waterborne disease, as does the introduction of change in drinking water treatment technology or food supply production. For example, efforts to address concerns over protozoan contamination of water and chlorine resistance have led to an increase in the use of ultraviolet light for compliance with water quality standards. An increased reliance on UV light treatment has subsequently raised concern over virus resistance, in particular, adenovirus (Nwachuku et al. 2005; Thurston-Enriquez et al. 2003). Therefore, changes in current applications must be carefully evaluated. Expansion of our food supply sources to include regions with deteriorated irrigation water quality and poor hygiene has been linked to foodborne outbreaks where water played a role in the transmission of disease. An example of this is the protozoan parasite *Cyclospora caryentensis*, which was imported into the U.S. on produce from developing countries (Mansfield and Gajadhar 2004).

Microbes are highly adaptable to environmental pressures and continue to evolve. The evolution of microbes can lead to a genetic reassortment that may increase the virulence of, or expand the host base for, that organism. A recent example of this is the severe acute respiratory syndrome (SARS) virus, which moves from the bat population to other animals and humans (Bennett 2006). The development of molecular methods for pathogen detection and source tracking has aided in the monitoring of water supplies and identifying causative agents in outbreak situations. Cultural methods are often necessary for the determination of pathogen viability in water, but for some of the most prevalent pathogenic microbes, laboratory methods for cultural detection have not been developed or standardized (i.e., noroviruses).

III. Agents of Waterborne Disease

Pathogens capable of causing waterborne or water-based illnesses include viruses, bacteria, and protozoa (Table 3). Also of concern in some geographical regions are helminths and blue-green algae. Since 1971, the

Table 3. Agents of Waterborne or Water-based Disease.

Category	Pathogen
Bacteria	*Vibrio cholerae*
	Salmonella spp.
	Shigella spp.
	Toxigenic *Escherichia coli*
	Campylobacter spp.
	Yersinia enterocolitica
	Plesiomonas shigelloides
	Legionella
	Helicobacter pylori
Protozoa	*Giardia lamblia*
	Cryptosporidium parvum
	Entamoeba histolitica
	Cyclospora cayetanensis
	Isospora belli
	Microsporidia
	Ballantidium coli
	Toxoplasma gondii
	Naegleria fowleri
Viruses	Norovirus
	Sapprovirus
	Poliovirus
	Coxsackievirus
	Echovirus
	Paraechovirus
	Enteroviruses 69-91
	Reovirus
	Adenovirus
	Hepatitis A
	Hepatitis E
	Rotavirus
	Astrovirus
	Picobirnavirus
	Coronavirus

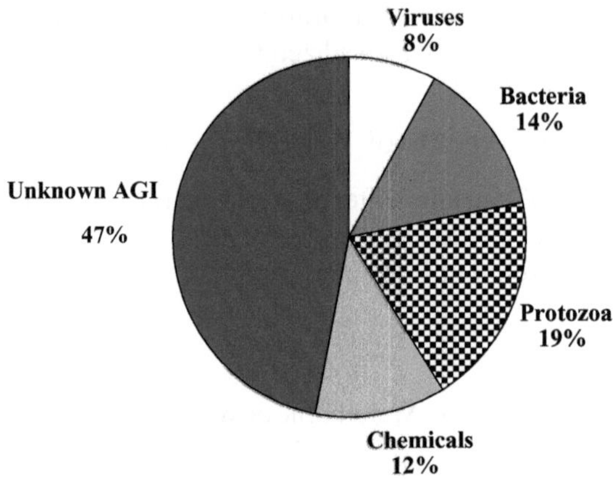

Fig. 1. Drinking water outbreaks by etiological agent, 1971–2002 ($n = 764$). AGI, acute gastrointestinal illness. (From Blackburn et al. 2004; Calderon 2004.)

Centers for Disease Control (CDC), USEPA, and other agencies have been collecting data regarding waterborne disease outbreaks in the U.S. From 1971 to 2002 there have been 764 documented waterborne outbreaks associated with drinking water, with 12% caused by chemicals, 14% by bacteria, 19% by protozoa, and 8% by viral pathogens (Fig. 1). Nearly half of all documented waterborne outbreaks since 1971 were caused by an undetermined etiology, i.e., acute gastrointestinal illness (AGI). The characteristics of these outbreaks of unknown AGI are often consistent with a viral etiology, some of which are known to be nonculturable. Outbreaks during 1971–2002 are known to have resulted in 575,457 cases of illness and 79 deaths; however, the true impact of waterborne disease is estimated to be much higher. For example, Morris and Levin (1995) estimate that 7 million people become ill and more than 1,000 die each year as a result waterborne microbial infections.

A. Viruses

Viruses range from 0.01 to 0.1 μm in size, are obligate, intracellular parasites, and are capable of long-term survival in the water environment (weeks to months). Viruses of greatest concern in water, and their associative illnesses, include enteroviruses (diarrhea, meningitis, myocarditis, fever, respiratory disease, nervous system disorders, birth defects), hepatitis A virus (hepatitis, liver damage), noroviruses (diarrhea), astrovirus (diarrhea), adenovirus (diarrhea, respiratory disease, eye infections, heart disease), and rotavirus (diarrhea).

Viruses have the greatest infectivity, requiring the fewest number to cause infection, of all waterborne microorganisms, are excreted in the feces in the largest numbers (up to 10^{11}/g), and generally have the longest survival in the environment; most only infect humans. They are not efficiently removed by conventional filtration and are more resistant to disinfectants than bacteria. Because of their small size and ease of transport in the subsurface, viruses are of primary concern in groundwater. Viruses are known to be the causative agent in 8% of drinking water outbreaks reported in recent years (Fig. 2).

B. Bacteria

Bacteria are prokaryotic, single-celled organisms surrounded by a membrane and cell wall, ranging in size from 0.1 to 10 μm. Enteric bacteria are able to colonize the human intestinal and gastrointestinal tract. Generally, enteric bacteria do not survive long in the environment, although some have resistant spores or can form dormant stages that aid in their survival. Waterborne outbreaks caused by enteric bacteria primarily occur because of failed or absent treatment processes. Examples of waterborne enteric bacteria include *Salmonella* (typhoid, diarrhea), *Shigella* (diarrhea), *Campylobacter* (diarrhea, nervous system disorders), *Vibrio cholerae* (diarrhea), and *Escherichia coli* (certain strains: diarrhea, hemorrhagic colitis). *Legionella* (pneumonia, respiratory infections) is an important water-based bacteria, and reports of *Legionella* outbreaks have only recently been added to the CDC surveillance summaries; however, six water-associated outbreaks were recorded in 2001–2002 (Blackburn et al. 2004). Non-*Legionella* bacteria are known to have caused 17% of the waterborne outbreaks documented from 1991 to 2002 (see Fig. 2).

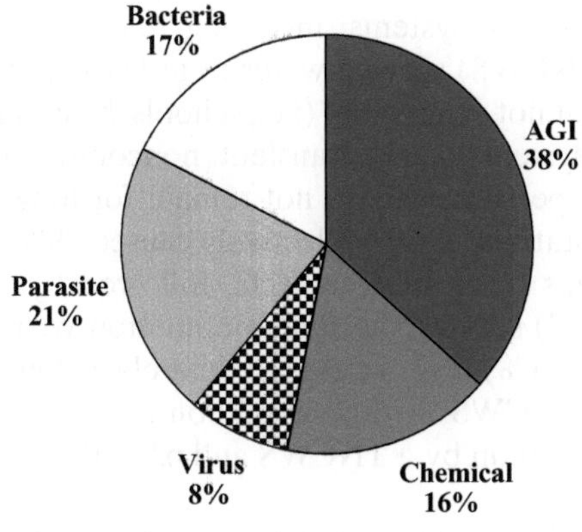

Fig. 2. Waterborne outbreaks associated with drinking water by etiological agent, 1991–2002 ($n = 183$). AGI, acute gastrointestinal illness. (From Barwick et al. 2000; Blackburn et al. 2004; CDC 1993; Kramer et al. 1996; Lee et al. 2002; Levy et al. 1998.)

Helicobacter pylori is a bacterium that has recently been recognized as the primary cause of duodenal (90%) and gastric ulcers (80%) (CDC 2001). It is considered a class A carcinogen, meaning that infections can lead to gastric cancer, the second most common cancer worldwide. Although the disease contribution related to the waterborne route of exposure is uncertain, studies have found 10%–60% of individual groundwater wells contaminated with *H. pylori* (Park et al. 2001).

C. Protozoa

Protozoan parasites are single-celled animals that live in the gastrointestinal tract of infected individuals. They range in size from 1 to 100 μm and produce an environmentally stable cyst or oocyst stage. The thick cyst or oocyst walls are highly resistant to disinfectants used in conventional water treatment. *Crytosporidium* and *Giardia lamblia*, both causing diarrhea, are the primary protozoa of concern with regard to water quality in the U.S. *Cyclospora caryentensis* is another parasite that has been linked to a possible waterborne outbreak in the U.S. (Mansfield and Gajadhar 2004). *Naegleria fowleri* is a water-based pathogen of primary concern because of a high fatality rate in diagnosed cases. Two deaths occurred in an outbreak of *Naegleria* in 2002 (Blackburn et al. 2004). Overall, protozoa caused 21% of drinking water outbreaks from 1991 to 2002 (see Fig. 2).

IV. Drinking Water Outbreaks

During the most recent 12-yr survey of waterborne disease (1991–2002), there were 183 documented outbreaks associated with drinking water. Most (76%) were from a groundwater source, with 18% linked to surface water systems (Fig. 3).

Public noncommunity systems, including nontransient noncommunity water systems (NTNCWS) serving water to at least 25 of the same people at least 6 mon/yr, but not year round (i.e., schools, hospitals, and offices with their own water systems), and transient noncommunity water systems (TNCWS), serving persons who do not remain for long time periods (i.e., campgrounds, gas stations, etc.), collectively caused 39% of drinking water-associated outbreaks from 1991 to 2002, followed by public community water systems (CWS) serving the same population year round (36%) and individual systems (25%) (Fig. 4). Approximately 264 million people in the U.S. are served by a CWS, with 19.8 million served by a noncommunity water source (12.9 million by a TNCWS and 6.9 million by a NTNCWS) in the U.S.

Although a drinking water outbreak was more likely to occur in a noncommunity supply utilizing a groundwater source, outbreaks involving the greatest number of individuals exposed occurred in CWS from a surface water source (Table 4).

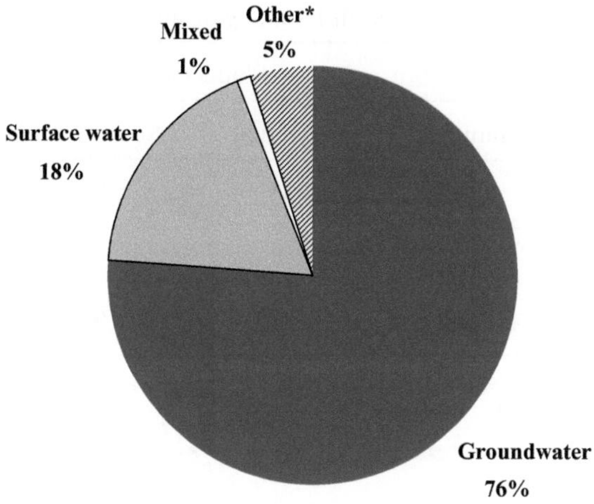

Fig. 3. Documented disease outbreaks associated with drinking water by source, 1991–2002 ($n = 183$). (From Barwick et al. 2000; Blackburn et al. 2004; CDC 1993; Kramer et al. 1996; Lee et al. 2002; Levy et al. 1998.)

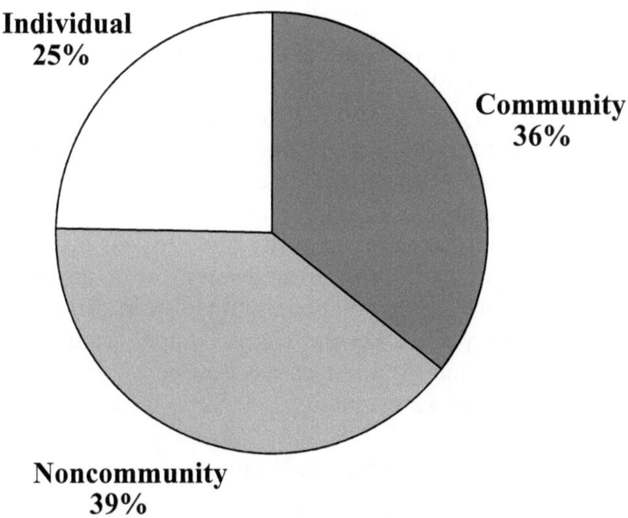

Fig. 4. Documented disease outbreaks associated with drinking water by system type, 1991–2002 ($n = 183$). (From Barwick et al. 2000; Blackburn et al. 2004; CDC 1993; Kramer et al. 1996; Lee et al. 2002; Levy et al. 1998.)

Table 4. Illness Cases Associated with Drinking Water Outbreaks, 1991–2002.

	Illness cases			
Source type	Community	Individual	Noncommunity	Total
Groundwater	3,967	364	9,468	**13,817**
Surface water	415,420	65	304	**415,789**
Mixed	3,013	0	0	**3,013**
Other	0	288	0	**288**
Total	**422,400**	**717**	**9,790**	**432,907**

Source: Barwick et al. (2000); Blackburn et al. (2004); CDC (1993); Kramer et al. (1996); Lee et al. (2002); Levy et al. (1998).

In the most recently published survey period (2001–2002), 23 of 25 (92%) outbreaks associated with drinking water were from a groundwater source, and 9 (39%) of these were associated with individual homeowner systems not regulated by the USEPA (Blackburn et al. 2004).

V. Sources of Microbial Contamination

Regarding pathogen exposure, contamination is not evenly distributed but rather affected by the number of pathogens in the source water, the age of the distribution system, the quality of the delivered water, and climatic events that can tax the treatment plant operations. Because it is not practical to monitor water supplies in real time and at the point-of-use for all groups of pathogens, episodic contamination events are difficult to predict or identify. From 1991 to 2002, the majority of outbreaks occurred because of a lack of treatment (primarily groundwater) or a treatment failure (Fig. 5). Efforts to control microbial contamination in drinking water are focused at four primary sites: (1) the source water, (2) treatment plant, (3) distribution system, and (4) point-of-use. Source water protection is the first step in control of the water quality.

A. Source Water

All surface waters, no matter how pristine, contain waterborne pathogens, because most of the significant waterborne pathogens are zoonoses, meaning they can be transmitted to humans from animals. Birds are a significant source of *Campylobacter*, as cattle are of *Cryptosporidium*. The more animal husbandry taking place near a watershed, the greater the concentration of zoonotic waterborne agents that can be expected in the water (Cox et al. 2005). Sewage discharges can also be a source of pathogens, even though they may be disinfected. Although chlorination is effective in reducing

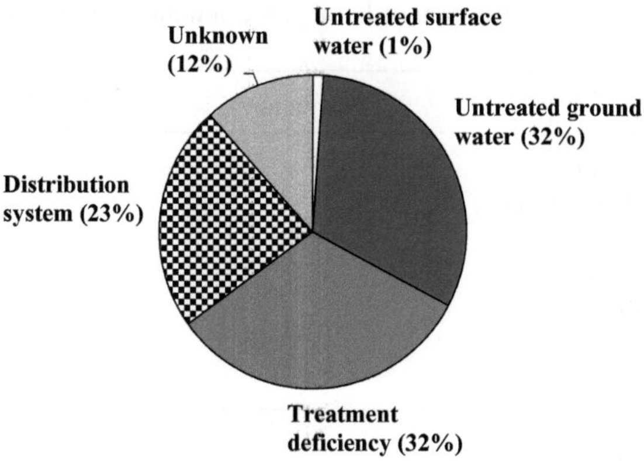

Fig. 5. Documented disease outbreaks associated with drinking water by deficiency, 1991–2002 ($n = 183$). (From Barwick et al. 2000; Blackburn et al. 2004; CDC 1993; Kramer et al. 1996; Lee et al. 2002; Levy et al. 1998.)

the number of bacterial pathogens, it has little effect on protozoan parasites and limited effectiveness on viral pathogens, as normally practiced (Fallacara et al. 2004). *Giardia* is more abundant in sewage discharges than *Cryptosporidium* (Smith and Grimason 2003). With the exception of hepatitis E virus, enteric viruses are not zoonotic and only originate from sewage sources, i.e., sewage treatment plants, combined sewer overflows, and septic tanks.

The occurrence of enteric pathogens in surface waters is highly variable, depending heavily on rainfall events. The elevated concentrations of pathogens after such events can pose a major challenge to water treatment plants. Extreme rainfall events and waterborne disease outbreaks from drinking water have been positively correlated in both the U.S. and Canada (Curriero et al. 2001; Thomas et al. 2006).

In contrast to surface waters, groundwater supplies were historically thought to be free of pathogenic microbes for reasons of the natural filtering ability of the subsurface environment and distance a microbe would have to travel to reach the groundwater source. Microbial contaminants that find their way into groundwater may originate as a result of lack of wastewater treatment or improper management of wastewater disposal, septic tank contamination, underground storage tank or landfill leaks, mismanagement of animal waste disposal, shallow wells, etc. Improved surveillance using molecular and cultural detection methods has led to increased evidence of human enteric viruses and other potentially harmful microbes in groundwater. Private groundwater wells are a concern because they are rarely, if ever, monitored and treated.

Table 5. Frequency and Duration of Pathogen Contamination Events in a Surface Drinking Water System.

Type of incident	Frequency[a]	Duration (hr/incident)
Suboptimal treatment:		
Wrong coagulant dosage	6	0.6
Filter operation	60	5
Chlorination failure	2	0.4
Pipe contamination:		
Cross-connection	0.00016	3 d
Main line	0.021	14 d

[a]Per 1,000,000 persons/y.
Source: Westrell et al. (2003).

B. Treatment

Much attention has been focused on enhancing current treatment processes to eliminate pathogens that are resistant to conventional water treatment, i.e., filtration for *Cryptosporidium*, and to expand treatment recommendations to source waters that were previously considered protected from harmful microbes, i.e., protected surface waters and groundwater. Populations are still at risk of pathogen exposure partly because of lack of treatment, i.e., no filtration of large municipal water supplies, such as New York and Boston, and currently no disinfection of municipal groundwater supplies. In addition, individual homeowners with private wells are at risk where contamination events would go largely unnoticed due to a lack of monitoring and reporting.

Even in the event of administering multibarrier treatment processes, it is not possible to remove 100% of the pathogens from the source water 100% of the time (Haas and Trussell 1998). Table 5 shows the documented frequency of various contamination events in a surface drinking water treatment system in Sweden, including treatment failures and distribution system contamination. Quantitative microbial risk assessment suggests that, depending on the original raw water quality, such events could cause serious health consequences (Westrell et al. 2003).

C. Distribution System

Even water that is adequately protected and treated is subject to pathogens entering the distribution system. From 1971 to 2002 there were 133 (17% of all outbreaks) documented waterborne outbreaks in the U.S. linked to distribution system contamination (Barwick et al. 2000; Blackburn et al. 2004; CDC 1993; Kramer et al. 1996; Lee et al. 2002; Levy et al. 1998). Preliminary data from the 2003–2004 survey period indicate that 38% of the reported outbreaks associated with drinking water systems were also associated with distribution systems (NRC 2006; Liang et al. 2006).

Municipal Distribution Systems

During the most recently published survey period (2001–2002), 5 of 25 (20%) of the documented waterborne outbreaks were associated with drinking water distribution system deficiencies and, of the 7 outbreaks reported involving community water systems, 4 (57.1%) were linked to distribution system problems (Blackburn et al. 2004). Although the overall number of reported outbreaks associated with community water systems has decreased in the last decade, the proportion of outbreaks associated with distribution systems has increased (Fig. 6). The reduction in total waterborne outbreaks is largely attributed to the promulgation of numerous regulations by the USEPA, including the surface water treatment rule, primarily aimed at reducing the risks of waterborne protozoa and improving water treatment (Pierson et al. 2001; Blackburn et al. 2004), but the current regulatory requirements do not appear to reduce the proportion of outbreaks associated with distribution systems.

The distribution system includes both the pumping, piping, and storage networks that deliver finished water to end users. There are approximately 1 million miles of distribution system networks in the U.S. and an estimated

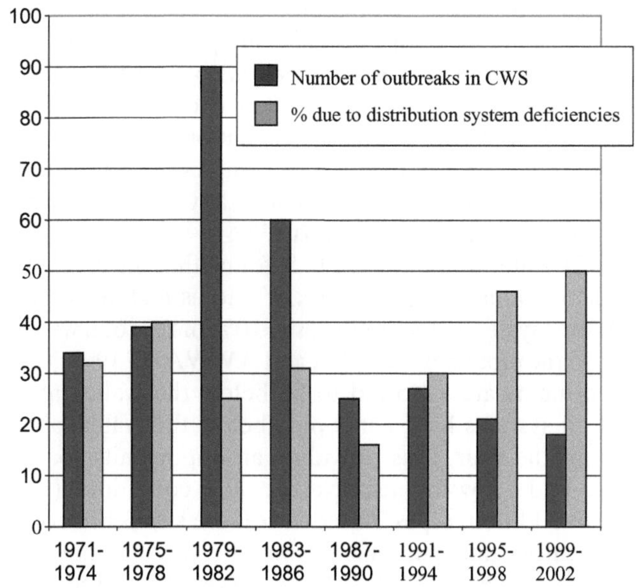

Fig. 6. Waterborne disease outbreaks in community water systems (CWS) associated with distribution system deficiencies. (Modified from NRC 2006.)

154,000 finished water storage facilities, with more than 13,000 miles of new pipes installed each year (AWWA 2003; Grigg 2005; Kirmeyer et al. 1994). In a 2005 report on the nation's infrastructure, the USEPA acknowledged the need for significant investment in installing, upgrading, or replacing infrastructure for delivering and storing drinking water at an estimated 20-yr cost of $208.4 billion (USEPA 2001, 2004). In the U.S. there is a wide range of distribution pipe age and materials with varying life expectancies. Pipes in the U.S. are replaced at an average rate of once every 200 yr (Grigg 2005); however, the life expectancies range from 75 to 120 yr (AWWA 2001; AWWSC 2002). Approximately 26% of the distribution pipes in the U.S. are in poor condition, and the annual number of documented main breaks has significantly increased from about 250 in 1970 to 2,200 in 1989 (AWWSC 2002). It is estimated that even well-run water distribution systems experience about 25–30 breaks per 100 miles of piping/yr (Deb et al. 1995). Using a value of 27 main breaks/100 miles/yr, Kirmeyer et al. (1994) estimated 237,000 main breaks/yr in the U.S.; however, variation between utilities is considerable. Haas (1999) reported results from a survey of water systems that showed a range of average main breaks of 488/yr for systems serving more than 500,000 people, to 1.33/yr for systems serving fewer than 500 people. The public health significance of these breaks in the distribution system is not currently known.

Maintaining the hydraulic integrity (positive pressure) of water distribution is important given that insufficient pressure has led to disease epidemics worldwide (reviewed in Lee and Schwab 2005). Negative hydraulic pressure creates a backflow of nonpotable water into the potable water supply via back-siphonage, where significant pressure drops siphon contaminants into the system at cross-connections or leakage points, or back-pressure from pressures in the system that exceed the supply pressure (Herrick 1997). Even minor pressure fluctuations create back-siphonage where intrusion rates are estimated at >1 gpm (LeChevallier et al. 2003a). During power outages, up to 90% of nodes have been shown to draw a negative pressure (LeChevallier et al. 2003b).

A survey of 26 water utilities in the U.S. found that the percent of leakage (unaccounted-for water) ranged from <10% to as high as 32% (Kirmeyer et al. 2001). Water systems commonly lose >10% of the total water produced through leaks in the pipelines (AWWA and AWWARF 1992). At least 20% of distribution mains are reported to be below the water table, but it is assumed that all systems have some pipe below the water table for some time throughout the year, thus providing an opportunity for intrusion of exterior water under low or negative pressure conditions (LeChevallier et al. 2003b). In addition, pipes buried in soil are subject to contamination with fecal indicators and pathogens from the surrounding environment (Karim et al. 2003; Kirmeyer et al. 2001). A survey of water utilities in North America found that 28.8% of cross-connections resulted in bacterial contamination (Lee et al. 2003). Negative hydraulic pressure can draw pathogens from the surrounding environment into the water supply where residual disinfection efficacy is uncertain and variable, depending on the magnitude of such events (Gadgil 1998; Haas et al. 1998; Trussell 1999). Little is known about the extensiveness of distribution system inadequacies and whether they are sporadic or continuously occurring (Lee and Schwab 2005), but outbreaks have been documented following external contamination in the distribution system despite the presence or requirement of residual disinfectant (Craun and Calderon 2001; Levy et al. 1998).

Decline in residual disinfectant is related to many factors, including the distance traveled, water flow velocity, residence time, age and material of pipes, and water pressure (Egorov et al. 2002). Although residual chlorine is present in the distribution system of treated water, the levels do not provide significant inactivation of pathogens in intrusion events (Payment 1999; Snead et al. 1980). More recent modeling studies have evaluated intrusion events at specific locations, with consideration to mixing, contact time, and other distribution system variables, before consumption. Under these realistic exposure scenarios, monochloramine disinfectants performed poorly against *Giardia* and *Escherichia coli*. Typical concentrations of chlorine residual (0.5 mg/L) inactivated *E. coli* in simulated sewage intrusion events but were again ineffective for *Giardia* (Baribeau et al. 2005; Propato and Uber 2004). Intentional contamination events in the distribution system are also a concern where public water supplies are potentially vulnerable to bioterrorism threats.

According to a Centers for Disease Control (CDC) survey, cross-connections and back-siphonage caused the majority (51%) of outbreaks linked to the distribution system from 1971 to 2000, followed by water main contamination (a collective 33%) and contamination of storage facilities (16%) (Fig. 7). Data compiled by the USEPA indicate that only a small percentage of contamination from cross-connections and back-siphonage are actually reported and that the CDC data underreports known instances of illnesses caused by backflow contamination events. For example, from 1981 to 1998, only 97 of 309 (31%) documented incidents were reported to public health authorities (USEPA 2002). Of the 97 reported incidences, 75 (77%) reported illnesses (4,416 estimated cases); however, only 26 (27%) appear in the CDC summaries of waterborne disease outbreaks.

Water quality may degrade in treated water storage facilities because of loss of disinfectant residual, increased temperature, and external contamination from birds, insects, animals, wind, rain, algae, etc. Storage tanks are particularly vulnerable to contamination in the absence, or failure, of a protective cover or barrier, open hatches, and vents; however, birds have been known to contaminate even covered public water supply distribution storage tanks (AWWA and EES 2002; Clark et al. 1996).

Home Distribution Systems

Bacterial colonization of pipes, connections, and faucets positioned along the channels of drinking water distribution, including the utility's

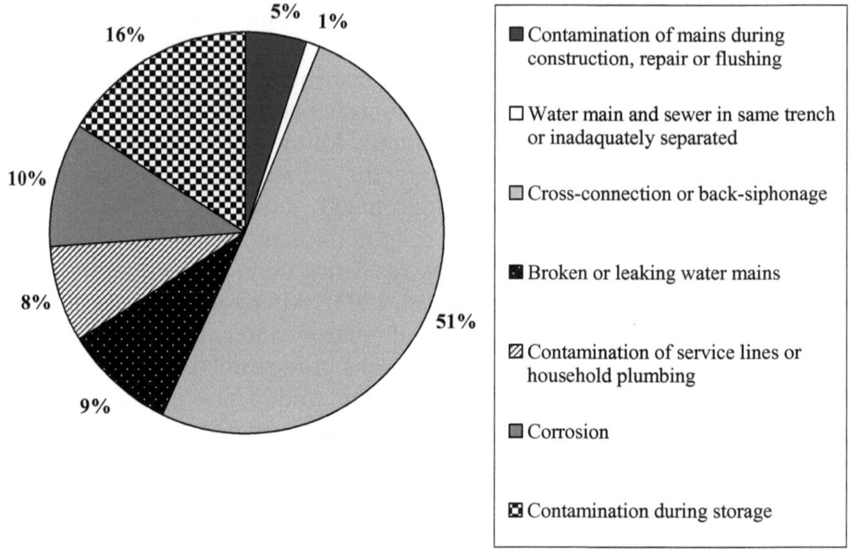

Fig. 7. Waterborne outbreaks caused by distribution system deficiencies, 1971–2000 ($n = 120$). (From Calderon 2004.)

Table 6. Tracking Deteriorating Water Quality to the Tap.

Sample site	HPC (cfu/mL)
Groundwater source	1–10
Distribution system	10–100
Household tap	1,000–1,000,000

Source: Pepper et al. (2004).

distribution system, the homeowner's premise plumbing, and fixtures in the home is well documented. Pepper et al. (2004) found that the bacteriological quality of water significantly deteriorates in the home plumbing relative to the distribution system, as evidenced by survey of heterotrophic plate count (HPC) bacteria (Table 6). Stagnant water in premise plumbing provides an environment where bacteria can grow to values several orders of magnitude higher than in the municipal distribution system (Edwards et al. 2005). Although HPC bacteria in drinking water is not considered a direct health risk (WHO/NSF 2003), opportunistic pathogens such as *Legionella* and *Mycobacterium* are associated with human disease and have been found in premise plumbing biofilms (Flannery et al. 2006; Pryor et al. 2004; Thomas et al. 2006; Tobin-D'Angelo et al. 2004; Vacrewijck et al. 2005).

Since 2001, *Legionella* outbreaks have been documented in the CDC surveillance summaries of waterborne disease and comprise a significant portion of drinking water outbreaks (19% in 2001–2002). All six of the documented *Legionella* outbreaks in 2001–2002 were related to regrowth of *Legionella* in the distribution systems of large buildings or institutions (Blackburn et al. 2004).

VI. Geographical Distribution of Reported Violations

Predicting the most at-risk populations based on geographical distribution is difficult for reasons of the relative significance of source water type and quality, treatment plant reliability, climatic events, distribution system integrity, reporting bias, and other factors. The Safe Drinking Water Information System provides data on CWS reporting health based violations of the National Primary Drinking Water Regulations (NPDWRs). The NPDWRs are legally enforceable standards that apply to public water systems subject to inorganic, organic, radionuclide, microbial, or other health-effecting contaminants. These primary standards set maximum contaminant levels (MCLs), the maximum permissible level of a contaminant in water delivered to any user of a public water system. In fiscal year 2003, 3,986 CWS (8% of total systems reporting) serving 24.4 million people (9% of the population) delivered drinking water in violation of at least one of the health-based standards (Fig. 8). Although about half of these violations

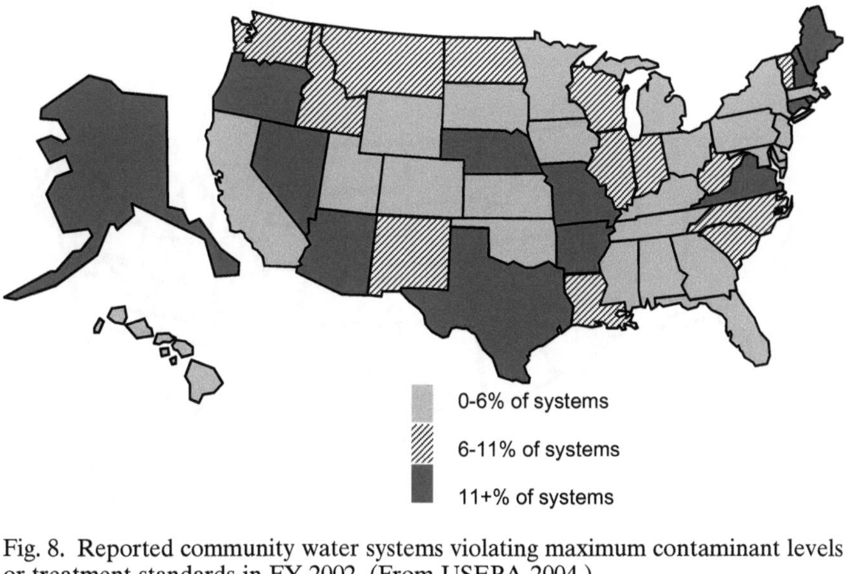

Fig. 8. Reported community water systems violating maximum contaminant levels or treatment standards in FY 2002. (From USEPA 2004.)

were the result of monitoring and reporting errors, the top two reported violations were under the category of the total coliform rule (9,056 reported violations) and the surface water treatment rule (1,747 reported violations). U.S. commonwealths and territories (i.e., American Samoa, Puerto Rico, U.S. Virgin Islands) were documented with an average of 44% of CWS reporting health-based violations, potentially impacting 71% of the population (USEPA 2004). Most of the U.S. population receives water from a CWS. Although there are 54,064 community water systems, serving a total of 263.9 million people, just 7% serve 81% of the population (USEPA 2006b).

VII. Evidence of Groundwater Vulnerability

Community water systems have more groundwater than surface water sources, but more people drink from a surface water system. A reported 11,403 systems, serving 178.1 million people, relied on surface water compared to 42,661 systems, serving 85.9 million people, reliant upon groundwater sources (USEPA 2006b). States with groundwater sources serving the greatest number of individuals are diagramed in Fig. 9. Before the newly promulgated Ground Water Rule (USEPA 2006), utilities with a groundwater source were not required to disinfect the water supply and many small communities and individual homeowners continue to consume untreated groundwater. Several national surveys have documented evidence of viruses in groundwater (Table 7). The newly promulgated Ground Water Rule applies to more than 147,000 public water systems and more

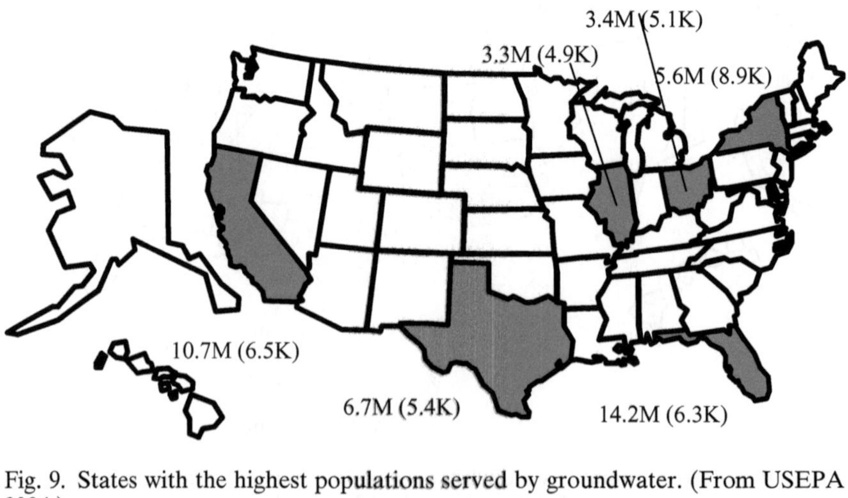

Fig. 9. States with the highest populations served by groundwater. (From USEPA 2004.)

Table 7. Evidence of Enteric Virus Contamination in U.S. Groundwater Wells.

Sample description	Virus positive	Source
448 utility wells, 35 states	32% enteric virus	Abbaszadegan et al. (2003)
50 homeowner wells	8% enteric virus	Borchardt et al. (2003a)
29 utility wells	16% enteric virus	Fout et al. (2003)
48 midwest utility wells	42% enterovirus 6% norovirus group 1	Borchardt et al. (2004)
211 Californian utility wells	10% enterovirus	Yates (unpublished, 2004)

than 100 million consumers, utilizing municipal groundwater sources. The rule requires that sanitary surveys be conducted by December 31, 2012, for most CWS and by 2014 for CWS with outstanding performance and for all noncommunity water systems, to help identify deficiencies that may lead to impaired water quality. Source water monitoring for indicator microbes, corrective actions for systems with significant deficiencies or source water fecal contamination, and compliance monitoring are further required. The USEPA estimates that the Ground Water Rule will reduce waterborne viral illnesses by approximately 42,000 cases each year, a 23% reduction from the current baseline estimate.

One survey of 448 utility wells in 35 states, using molecular methods of detection (reverse transcriptase-polymerase chain reaction, RT-PCR), found evidence of enteric viruses, including enterovirus, rotavirus, and hepatitis A RNA, in approximately 32% of groundwater supplies (Abbaszadegan et al. 2003). Molecular methods for virus detection do not determine viability, and thus the public health significance of these results is not known, but the presence of viral RNA in groundwater suggests a potential for exposure and adverse health risks.

An additional survey of 321 samples from 29 U.S. utility wells, collected over 1 yr, detected human enteric viruses including enterovirus, reovirus, norovirus, and hepatitis A virus in 72% of the sites and 16% of the samples using RT-PCR (Fout et al. 2003). Similarly, 50% of samples from 48 midwest utility wells tested positive for human viruses (Borchardt et al. 2004). In the latter study, three samples were found positive for culturable hepatitis A virus. Another study of 50 private homeowner wells found enteric viruses in 8% of the samples collected (Borchardt et al. 2003a). In addition to human viruses, protozoan parasites have been documented in groundwater. Of 199 groundwater samples surveyed, 5% of vertical wells, 20% of springs, 50% of infiltration galleries, and 45% of horizontal wells tested positive for *Cryptosporidium* oocysts, calling for a reevaluation of the notion that groundwater is inherently free of protozoan parasites (Hancock et al. 1998).

Helicobacter pylori has been found in biofilms of water distribution systems (Park et al. 2001) and individual groundwater wells. Epidemiological studies in Germany have linked infection in children with drinking untreated well water serving individual homes (Herbarth et al. 2001), as did a study in West Virginia linking contaminated homeowner wells (Elitsur et al. 1998). Studies of groundwater quality have implicated an association with septic systems and disease. Borchardt et al. (2003b) found that viral diarrhea in children from 14 contiguous zip codes in Wisconsin positively correlated with septic tank density. Water holding tanks and bacterial diarrhea were also positively correlated. Raina et al. (1999) showed *E. coli* in well water was correlated to diarrhea in rural families. The closer the septic system was in proximity to the drinking water well, the greater the incidence of disease. Overall, 46% of wells were contaminated if the septic system was within 20m.

VIII. Estimating Waterborne Disease risk in the United States

Variable approaches have been used to estimate gastrointestinal illness from waterborne pathogens including epidemiological studies and exposure analysis. Information is lacking, however, regarding risk estimates considering gastroenteritis and other illnesses related to microbial contaminants in drinking water.

A. Estimates of Gastroenteritis from Epidemiological Studies

Estimating the incidence of endemic acute gastrointestinal illness attributable to drinking water has been approached using information obtained from household intervention trials (Colford et al. 2002, 2005; Hellard et al. 2001; Payment et al. 1991, 1997). Determining the illness attributable to drinking water involves estimating the baseline of gastrointestinal illness within communities, and such information can be useful when conducting quantitative microbial risk assessments of drinking water quality. The household intervention trials conducted in the research investigations cited above are types of epidemiological studies that involve randomly designating one group of households as the "intervention group" where household members utilize drinking water obtained via an in-home treatment system and then having another group of households use water directly from their tap or through a fake device that provides no additional water treatment. In the latter situation, the study is *blinded*, meaning that neither group knows during the study whether their in-home device is actually providing treatment.

Such household intervention trials have been conducted in the U.S., Canada, and Australia. Table 8 lists and highlights various aspects of these studies. For all these trials, the human health outcome of interest was gastrointestinal illness, with some variation regarding the specific symptoms in defining that outcome. All participants were immunocompetent individuals who kept health diaries throughout the study to record symptoms related

Table 8. Household Drinking Water Intervention Trials Addressing Gastrointestinal Illness.

Study	Payment et al. (1991)	Payment et al. (1997)	Hellard et al. (2001)	Colford et al. (2002)	Colford et al. (2005)
Location	Montreal, Canada	Montreal, Canada	Melbourne, Australia	California, United States	Iowa, United States
Sample size (individuals)	2,408	5,253	2,811	236	1,296
Treatment via intervention	Reverse osmosis	Tap water with/without purge valve Bottled plant water and bottled purified water	Ultraviolet and 1-µm filter	Ultraviolet and 1-µm filter	Ultraviolet and 1-µm filter
Sham device?	No	No	Yes	Yes	Yes
Means of assessing illness	Health diary	Health diary	Health diary	Health diary	Health diary
Source water	Surface water (river) w/fecal coliforms and viruses	Surface water (river) w/protozoa and viruses	Surface water from catchments	Surface water w/*Cryptosporidium*	Surface water (river) w/indicators of fecal contamination and protozoa
Source water treatment applied	Combination of flocculation, filtration, ozonation, and chlorination	Combination of flocculation, settling, filtration, ozonation, and chlorination	Chlorination	Conventional treatment	Conventional treatment
Finished water quality	Met standards	Met standards	No total coliforms detected	Met standards	Met standards

Table 8. (*cont.*)

Study	Payment et al. (1991)	Payment et al. (1997)	Hellard et al. (2001)	Colford et al. (2002)	Colford et al. (2005)
Distribution system water quality	Not reported	No report of fecal contamination	Total coliforms and relatively high heterotrophic plate count bacteria detected in some samples	Not reported	Met standards
Attributable risk	0.26	0.08 for tap water 0.12 for tap water w/purge valve 0.02 for bottled plant water	0.03	0.85	Two periods: 0.02 (I) 0.14 (II)
Cases attributable to tap water	34%	12% for tap water 17% for tap water w/purge valve 3% for bottled plant water	4%	24%	0.008% (I) 0.08% (II)

Source: Modified from Colford et al. (2006).

to gastrointestinal illnesses. The source (surface) waters were reported to have varying levels of microbial contamination.

Payment et al. (1991) were first to conduct a household intervention trial addressing (gastrointestinal) illness attributable to drinking tap water. The tap water met both Canadian and U.S. regulations, but the source water was subject to contamination from sewage. The limitation of this study is that it was not blinded, so those persons drinking tap water (therefore no treatment device) may have been more inclined to report poorer health symptoms. An overall conclusion from this study is that an estimated 35% of the gastrointestinal illnesses occurring within the tap water group may be attributable to their drinking water.

Payment et al. (1997) conducted a follow-up study to address the results from their previous study (Payment et al. 1991). A goal of this investigation was to evaluate the role of distribution system water quality in gastrointestinal incidence, which resulted in a study design involving four groups of participants: a tap water group and a bottled purified water group (to address those exposed and unexposed, respectively), and a plant bottled water group and a tap water group using a purge valve (to address distribution system water quality). The attributable risk percent ranged from 3% for the bottled plant water group (rate of illness for this group is the same as for those drinking the bottled purified water), to 12% for the tap water group, to 17% for those in the tap water group with a purge valve. The investigators concluded that the excess number of gastrointestinal illnesses observed in the first study may not have been associated with surface water contamination but rather was associated with contamination within the distribution system because the rate of illness of the bottled plant water group was similar to those drinking bottled purified water. A limitation, however, is that about half the participants in the bottled plant water group dropped out during the course of the study. In addition, as with their previous study (Payment et al. 1991), this study was also unblinded.

Hellard et al. (2001) designed the first blinded household intervention study, which was conducted in Australia. Some participants used a water treatment device that involved an ultraviolet application and filtration while others were given a fake (no treatment) device. As with the Payment et al. studies (1991, 1997), participants recorded gastrointestinal illness symptoms in diaries, although this study used a slightly more strict definition of gastrointestinal illness. Participants were followed for more than 1 yr, another strength of the study besides the blinding of participants, and the investigators observed similar rates of illness of the participants using the fake device as those using the water treatment device.

Colford et al. (2002, 2005) conducted two household intervention trials in the U.S. Their first study was designed as a pilot to obtain information regarding the practicality of a study design to include blinding of participants. The investigators utilized a blinding index (James et al. 1996), which led to the conclusion that the study they designed could incorporate effective participant blinding. The investigators observed an attributable risk of 0.85 and concluded that 24% of the gastrointestinal illnesses could be attributable to tap water.

The goal of the Colford et al. (2005) follow-up study was to determine if water treatment at the tap could reduce the number of gastrointestinal illnesses. Again, the investigators used a blinding approach, and participants recorded gastrointestinal health symptoms in diaries. The investigators observed no difference in the rate of illness between those participants using the fake device and those using the treatment device. The authors offered the explanation that perhaps their study owed this conclusion to successful water treatment practices and a well-maintained water distribution system. In addition, it was recognized that water consumption by the participants outside the home may have had some effect on the study results.

Colford et al. (2006) reviewed the household intervention trials described above as well as presenting an approach for estimating the occurrence of acute gastrointestinal illness in the U.S. that can be attributable to drinking water. Their proposed approach considers the following: (a) the estimated incidence of acute gastrointestinal illness in the U.S. of 0.65 episodes per person-year based on data collected from the Foodborne Diseases Active Surveillance Network (FoodNet) (Hawkins et al. 2002; Jones et al., 2007); (b) the attributable risks determined from the household intervention trials (median attributable risk of 0.08 and a median attributable risk percent of 12%); (c) the proportion of risks of acute gastrointestinal illness associated with source water and/or water treatment quality; (d) the number of people in the U.S. served by community water systems and consuming drinking water from surface water sources and groundwater sources; and (e) the number of people in the U.S. served by community water systems that are known to have either poor quality source water or poor water treatment.

Based on considerations just listed and various assumptions, Colford et al. (2006) estimated that as many as 11.69 million cases of acute gastrointestinal illness, occurring each year, may be attributable to drinking tap water in the U.S. Assumptions include the applicability of the attributable risk percent estimates from the household intervention trials to the entire U.S. population. In addition, the authors created scenarios assuming different risk levels associated with either poor source water quality/poor water treatment or problems with quality within a distribution system. The latter resulted in the 11.69 million cases of acute gastrointestinal illnesses/yr estimation and a lower estimate of 4.26 million cases/yr associated with poor source water quality/poor water treatment (Colford et al. 2006). The authors emphasize that the primary purpose of their estimation of acute gastrointestinal illness incidence attributable to drinking tap water in the U.S. is to demonstrate a methodology that can be improved upon with more data.

Besides household intervention trials, community intervention studies have also been conducted to address waterborne gastrointestinal disease risks (Calderon 2001; Frost et al. 2006; Goh et al. 2005; Hellard et al. 2002; Kunde et al. 2006; McConnell et al. 2001). These types of studies offer some advantages over household intervention trials including that they may be simpler and less costly to conduct (Calderon and Craun 2006): they have included cohort, case-control, and ecological types of designs. Two of these (Calderon 2001; Goh et al. 2005) concluded that a reduction in gastrointestinal illnesses was observed as a result of additional water treatment. A preliminary report from the Kunde et al. (2006) study also indicates a decrease in diarrheal illness risk in participants over age 35 following the intervention. Conversely, preliminary data analysis from the Frost study (2006) does not indicate a significant difference; however, analysis is reported to be ongoing for both of these aforementioned studies (reviewed in Calderon and Craun 2006).

B. Estimates of Gastroenteritis from Exposure

Messner et al. (2006) described an approach for estimating the incidence of gastrointestinal disease in the U.S. from drinking water. These investigators assume that for each population served by a community water system, a distribution of incidence rates of acute gastroenteritis can be estimated that can then be used to derive an overall national estimate of this disease attributable to drinking water in the U.S. They emphasize the need for addressing "mixtures" of pathogens as opposed to considering health risks from exposure to an individual pathogen as one of the premises for the approach described in this paper. The authors speculate that the mean incidence of acute gastrointestinal illness attributable to drinking water among community water systems ranges widely because of variations in source water quality, water treatment efficiencies, water quality within a distribution system, and water quality management practices.

Messner et al. (2006) propose the development of a "risk matrix" to categorize community water systems (CWS) based on relative microbial risk levels. The authors suggest connecting the information obtained from epidemiological studies regarding the incidence rate of acute gastrointestinal illness to risk factors identified in the epidemiological studies that have been conducted and to other CWS. Identification of these risk factors will allow for risk-based categorizing of other CWS that have similar characteristics, therefore assuming that generalizations can be made regarding all U.S. CWS and the populations they serve. Completing this process involves overcoming several challenges from the lack of data related to both pathogen occurrence and variation in survivability and infectivity, as well as knowing the actual efficiency of water treatment applications, as opposed to theoretical information. Messner et al. (2006) utilize in their approach specific information obtained during the Payment et al. studies (1991, 1997) regarding factors associated with source water/water treatment quality and factors related to distribution system deficiencies, and also therefore utilize the specific definition of gastrointestinal illness as defined in these studies, for highly credible gastrointestinal illness.

To estimate the incidence of acute gastrointestinal illness in the U.S. caused by drinking water, Messner et al. (2006) selected 2004 as their reference year and assumed that a certain number of cases are the result of source water/water treatment quality and a certain number are caused by distribution system deficiencies. In addition, the investigators assume a lognormal distribution to address the variability of relative microbial risk related to both, leading to an estimation of a statistical distribution of acute gastrointestinal illness among CWS within the U.S.. The authors consider that there is an approximate 5-log range regarding mean pathogen concentrations in source waters and a 2- to 6-log range regarding mean pathogen reduction. They used Monte Carlo simulations to ultimately compute this estimate of the distribution of acute gastrointestinal illness.

Based on the described assumptions, Messner et al. (2006) estimate that the mean national estimate of gastrointestinal illness, using the Payment et al. (1991, 1997) definition of highly credible gastrointestinal illness, attributable to drinking water is 0.11 cases/person/yr (with a 95% credible bound of 0.03–0.22) (Table 9). The investigators relate their 0.11 cases/person/yr estimate to the reported rate of diarrheal "episodes" of 1.3/person/yr (Imhoff et al. 2004) and estimate that the percentage of "episodes" attributable to drinking water is 8.5%. If this same percentage is assumed and applied to the Imhoff et al. (2004) reported incidence, due to all causes, for acute gastrointestinal illness of 0.72 cases/person/yr, the investigators estimate that the incidence of acute gastrointestinal illness attributable to drinking water is 0.06 cases/person/yr (95% credible interval of 0.02–0.12). When applying the 0.72 cases of acute gastrointestinal illness/person/yr

Table 9. Estimates of Gastrointestinal Illness Attributable to Drinking Water.

Health outcome and drinking water attribute	Mean illness incidence (cases/person-yr)	95% credible bounds
Total highly credible gastrointestinal illness[a]	0.11	0.03, 0.22
Due to water source/ treatment	0.048	0.011, 0.086
Due to distribution system	0.062	0.005, 0.16
Total acute gastrointestinal illness	0.06	0.02, 0.12
Due to water source/ treatment	0.03	0.006, 0.05

[a]Estimated using data from Payment et al. (1991, 1997) studies.
Source: Modified from Messner et al. (2006).

from all causes reported in Imhoff et al. (2004) to the 272.5 million people served by CWS (based on data in USEPA 2006b), this results in an estimate of approximately 196 million cases/yr of total acute gastrointestinal illness. Assuming, based on estimates provided above, that 8.5% of the cases are attributable to drinking water, this translates into approximately 16 million cases of acute gastrointestinal illness/yr, which is a little higher than the upper end estimate computed by Colford et al. (2006) (11.69 million cases/yr).

C. Estimates of Waterborne Disease from Exposure

The following section represents our estimates of waterborne infection and illness risks in the U.S. categorized by source water type (Fig. 10). These estimates are based on the total number of water systems in the U.S. and total populations exposed. Illness risk estimates represent all possible illnesses associated with the microbial infection, not only gastroenteritis.

Groundwater risk estimates are based on predicted number of viral infections and illnesses associated with viruses in groundwater. For both the community system and noncommunity system groundwater risks, it was assumed that 10% of wells are positive for infectious viruses, assuming one infectious virus for every positive well. Dose–response data for rotavirus (Gerba et al. 1996b) were used, an exposure volume of 1.4 L/person/d was assumed (Covello and Merkhofer 1993) and the yearly risk was calculated based on a 350-d exposure (Aboytes et al. 2004).

When considering the number of people served by groundwater supplies, the number of viral infections estimated is 10.7 million/yr and 2.2 million/yr for community and noncommunity systems, respectively. Assuming that half of all infections lead to illness (Haas et al. 1993), this results in 5.4 million cases/yr associated with community groundwater systems and 1.1 million cases/yr associated with noncommunity groundwater systems. These infection and illness estimates may offer higher estimates of risk due to the high infectivity associated with rotavirus exposure. In addition, this exercise represents a methodology for estimating risks associated with exposure to any waterborne virus and infections that may lead to a wide range of clinical outcomes, not only gastroenteritis.

The estimated number of infections and illnesses associated with exposure to pathogens in municipal surface waters were also determined. This exercise utilized the same assumptions described for the groundwater risks as well as dose–response data for *Cryptosporidium* (exponential model; Messner et al. 2001) and *Campylobacter* (beta-Poisson model; Medema et al. 1996). Assuming a 1% frequency of contamination among the more than 11,000 CWS in the U.S. using surface water, and combining risk estimates associated with *Cryptosporidium*, *Campylobacter*, and rotavirus, 26 million *infections*/yr are estimated. When assuming that 50% of these infections will result in some type of illness, 13 million illnesses/yr are predicted.

Risk calculation 1: Estimated number of viral infections and illnesses associated with community groundwater systems

- Assume
  - Of 42,661 groundwater systems serving 86 million people (USEPA 2006b), 10% contain infectious virus
  - One infectious virus per liter (100-L samples analyzed)
  - 1.4 L water consumed per day (Covello and Merkhofer 1993)
  - 50% infections result in illness (Haas et al. 1993)
- Annual risk of infection = 0.12

- **10.7 million infections per year**
- **5.4 million illnesses per year**

Risk calculation 2: Estimated number of viral infections and illnesses associated with noncommunity groundwater supplies

- Assume
  - Of 111,036 groundwater systems serving 18 million people (USEPA 2006b), 10% contain infectious virus
  - One infectious virus per liter (100-L samples analyzed)
  - 1.4 L water consumed per day (Covello and Merkhofer 1993)
  - 50% infections result in illness (Haas et al. 1993)
- Annual risk of infection = 0.12

- **2.2 million infections per year**
- **1.1 million illnesses per year**

Risk calculation 3: Estimated number of infections and illnesses associated with municipal surface water supplies

- 178,000,000 persons supplied by surface water supplies in the U.S. (USEPA 2006b)
- Combined risk of *Cryptosporidium*, *Campylobacter*, and rotavirus infections
- Based on 1% frequency of contamination events
- Assuming 50% infections result in illness (Haas et al. 1993)

- **26.0 million infections per year**
- **13.0 million illnesses per year**

Fig. 10. Estimates of waterborne infection and illness risks in the U.S. categorized by source water type.

Total estimated number of waterborne illnesses per year in the U.S.

- Groundwater (municipal) = 5,400,000
- Groundwater (noncommunity) = 1,100,000
- Surface water supplies = 13,000,000

Total estimate = 19,500,000

Risk of viral infection associated with exposure to contaminated groundwater using rotavirus dose–response data

- beta-Poisson model for rotavirus (Gerba et al. 1996b):

  Probability of infection (P_i) $= 1 - (1 + N/\beta)^{-\alpha}$
  where $\alpha = 0.26$ and $\beta = 0.42$

- Probability of infection for 1 virus ($n = 1$): **0.27**

- Concentration of viruses in groundwater:
  - Assuming 4,266 positives/total volume analyzed (100 L per sample)
  - Assuming one virus per positive sample and 10% are positive = **0.001**

- Daily risk:
  [Concentration (0.001 viruses/L)][P_i for 1 virus (0.27)][1.4 L/d ingestion]
  = **0.000378 infections**

- Annual risk:
  $1 - (1 - \text{daily risk})^{350}$ = **0.12 infections per year**

Fig. 10. (*cont.*)

When combining illness estimations from all water sources addressed in this exercise, the total is 19.5 million cases/yr in the U.S. associated with drinking water. This estimation is higher than both the Colford et al. (2006) illness estimate, upper estimate of almost 12 million cases/yr, and the Messner et al. (2006) illness estimate, 16 million cases/yr, yet our estimate potentially reflects all health outcomes associated with exposure to pathogens in drinking water rather than just gastrointestinal illness.

IX. Water Treatment at the Point-of-Use

Water treatment technologies at the point-of-use can provide an additional barrier of protection from waterborne contaminants, particularly those entering the distribution system and present in premise plumbing. Point-of-use (POU) water treatment devices may be installed at the end of the faucet, plumbed in-line, or stand-alone pitchers, or they may be point-of-entry (POE) systems installed where water from the distribution system enters the premise plumbing. Many POU/POE systems are designed for aesthetic (i.e., taste, odor, hardness) improvements only, while others employ technologies to remove organic and inorganic chemicals, pathogens, bacteria, and radionuclides. According to a survey by the Water Quality Association (2001), 41% of homes in the U.S. report having a water treatment device in place at the point-of-use or the point-of-entry and 39% drink bottled water (Fig. 11). Most report the use of a tabletop pitcher, which are currently not designed or marketed for eliminating microbial pathogens from drinking water. Generally, systems designed to eliminate a wide variety of physical, chemical, and biological contaminants are costly and require routine professional maintenance. Membranes used for filtration in POU/POE devices must be changed at regular intervals, and systems with ultraviolet light disinfection must be routinely inspected for buildup on the lamps that could prevent effective light emission. Improper maintenance of

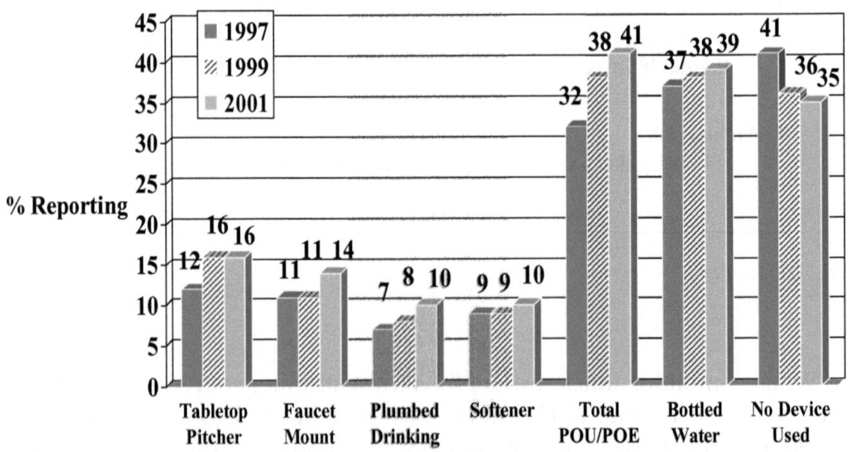

Fig. 11. Use of point-of-use (POU) water treatment devices in the U.S., 1997–2001 (WQA 2001.)

POU/POE treatment systems could result in exposure to a greater concentration of pathogens. In addition, heterotrophic plate count (HPC) bacteria often increase by several orders of magnitude in POU/POE water treatment devices.

A. Significance of Regrowth in POU/POE Water Treatment Devices

Several factors are related to bacterial regrowth in water, including, filtration, temperature, disinfectant type and residual, assimilable organic carbon level, corrosion control, and distribution pipe material. HPC bacteria are able to persist and grow in and on point-of-use/entry treatment device media, membranes, filters, and other surfaces to concentrations 10 fold or more higher in effluent waters. Granulated activated carbon (GAC) is a common medium used in POU/POE treatment devices known to support growth of HPC bacteria. Among other contaminants, GAC removes chlorine disinfectant residuals from tap water. Although this is desirable to improve the taste and odor of drinking water, the lack of disinfectant residual and the collection of bacterial growth substrates provides a suitable environment for HPC bacteria to attach to the media and grow, especially following periods of non-use and stagnation. HPC bacteria may also grow in water storage vessels, distribution pipes, pressure tanks, and hot water heaters.

Although many studies have documented the presence of large numbers of HPC bacteria in POU-treated water, there has been no correlation to increased disease (Allen et al. 2004; Calderon 1991; Colford et al. 2002; Edberg and Allen 2004; WHO/NSF 2003; Payment et al. 1991, 1997). Relative to HPC levels in common foods, water plays a minor role as a source of ingested bacteria (Stine et al. 2005). Certain HPC bacteria, such as *Pseudomonas*, *Klebsiella*, and *Aeromonas*, are opportunistic pathogens, meaning they are capable of causing disease in an immunocompromised host. Although these organisms can be isolated from treated water systems (Chiadez and Gerba 2004), ingestion is not their route of disease transmission. There is insufficient evidence linking these opportunistic pathogens to disease transmission via drinking water (Allen et al. 2004).

Several studies have shown that HPC bacteria in POU/POE treatement devices can outcompete human pathogens and may offer a protective effect to consumers (Camper et al. 1985; Gerba 2003; Rollinger and Dott 1987). Gerba (2003) found commercially available POU carbon filter devices placed on home faucets and used for 3–6 wk established a background culture of HPC bacteria within the systems. *Salmonella typhimurium*, *E. coli*, poliovirus, and hepatitis A virus, all known human enteric pathogens, were added to sterile tap water, regular tap water, and POU-treated tap water that was high in HPC organisms. HPC bacteria in the POU-treated water were clearly antagonistic to the pathogenic bacteria used in this study, reducing their counts >10 fold in 1 d and >10,000 fold in 2 d. A similar, but less dramatic, trend was seen with the pathogenic viruses.

Other studies supporting the antagonistic effect of HPC bacteria on pathogens have been conducted (Camper et al. 1985). Three enteric bacterial pathogens, *Yersinia enterocolitica*, *Salmonella typhimurium*, and enterotoxigenic *E. coli*, readily grew on sterile GAC; however, in the presence of water containing populations of HPC organisms, the pathogen counts gradually decreased. The most dramatic results were seen when bacterial populations from river water were previously established on GAC and a mixture of HPC and pathogenic bacteria were added to the media. Pathogens not only decreased at a more rapid rate but were prevented from initial attachment compared to sterile GAC filters. These studies suggest an antagonistic effect on pathogenic bacteria caused by the presence of HPC bacteria on the filters, possibly because pathogenic bacteria do not compete well in the presence of high HPC bacteria.

B. Health Benefits of POU/POE Water Treatment Devices

Few studies have directly targeted the benefits of POU water treatment systems for reduction of waterborne disease. Most of the available data are from epidemiological studies with a few incidental pieces of information from outbreak events. For example, survey data showed that no one who died in the waterborne *Cryptosporidium* outbreak in Milwaukee was using any type of fine filtration device for water treatment in the home (WQA 2002). In the same outbreak, persons who did have a point-of-use filtration device in place reported significantly lower incidences of diarrhea compared to those without (Addiss et al. 1996).

Epidemiological studies, in Canada, by Payment et al. (1991, 1997) suggest that 35% of all gastrointestinal illnesses could be waterborne when source water quality was degraded (see previous discussion). The 1997 study also found that children gain the most by having a POU water treatment system in place. In 2- to 5-yr-old children, drinking tap water resulted in an excess of 40% of gastrointestinal illness compared to those drinking tap water filtered at the point-of-use, and an excess of 17% was seen with children drinking bottled tap water versus POU-treated water.

Two epidemiological studies using randomized, blinded, controlled trials to evaluate risks related to tap water consumption determined that the risks were equal among groups supplied with POU-treated (1-μm filtration and UV disinfection chamber) water compared to untreated tap water (Colford et al. 2005; Hellard et al. 2001). Some of the uncertainties of these studies are that only a single water system was evaluated and, in the Hellard study, a pristine water source. The Colford study evaluated the Iowa-American Water Company, reported to be one of the best in the country, utilizing conventional filtration and a combination of chlorine and chloramine disinfectants. The study included intensive monitoring of the distribution system water quality and pressures and indicated high-quality delivery of the finished product.

More studies are needed to assess the impact of POU filtration systems for waterborne disease reduction. Future studies should evaluate multiple water systems over a wide geographical area to determine the efficacy of POU water filtration systems over varying finished water qualities. Direct monitoring of POU filters, placed in residential or commercial applications in regions where source water quality or distribution system integrity is questionable, would provide much-needed pathogen occurrence and exposure data.

X. Conclusions

Although current protocols in municipal treatment requirements are effective at eliminating pathogens from water if properly applied, inadequate, interrupted, or intermittent treatment has repeatedly been associated with waterborne disease outbreaks. Factors to consider with regard to pathogen exposure is that contamination is not evenly distributed but rather affected by the number of pathogens in the source water, the age of the distribution system, the quality of the delivered water, and climatic events that can tax the treatment plant operations.

Weather events are difficult to predict but are known to influence exposures to microbial pathogens by their increased transport and dissemination via rainfall and runoff and the survival and/or growth through temperature changes (CGER 2001). Effects of increased rainfall on watershed protection, infrastructure, and storm drainage systems affected by increased rainfall may lead to increased risk of contamination events. Extreme precipitation events preceded 51% of outbreaks from 1948 to 1994 (Curriero et al. 2001).

Current regulatory standards and monitoring requirements do not guarantee the absence of human pathogens in tap water. For example, the Total Coliform Rule, mandating the use of bacterial indicators of water quality, does not predict vulnerability to an outbreak (Craun et al. 2002). In fact, few community and noncommunity water systems that reported an outbreak from the survey period 1991–1998 had violated the coliform standard in the 12-mon period before the outbreak.

In 2002, the USEPA reported that 94% of the U.S. population served by community water systems received drinking water that met all applicable health-based drinking water standards through treatment and source water protection. An internal audit indicated that the figure was believed to be closer to 81% (USEPA 2004). Furthermore, little is known about exposures to waterborne pathogens in populations not served by public water systems where there is a general lack of monitoring.

Finally, of particular concern are sensitive populations in the U.S. that are susceptible to higher rates of infections and to more serious health outcomes from waterborne pathogens. These subpopulations include not only individuals experiencing adverse health status, but also those experiencing "normal" life stages, e.g., pregnancy, or those very young or old.

Individuals in any of these situations may want to consider additional strategies to prevent waterborne illness attributable to drinking water, such as the utilization of a point-of-use water treatment device. Better communication between water quality professionals and healthcare providers is needed to develop and distribute materials to inform the public of mitigation options beyond the current multibarrier approach of municipal water treatment.

Summary

Outbreaks of disease attributable to drinking water are not common in the U.S., but they do still occur and can lead to serious acute, chronic, or sometimes fatal health consequences, particularly in sensitive and immunocompromised populations. From 1971 to 2002, there were 764 documented waterborne outbreaks associated with drinking water, resulting in 575,457 cases of illness and 79 deaths (Blackburn et al. 2004; Calderon 2004); however, the true impact of disease is estimated to be much higher. If properly applied, current protocols in municipal water treatment are effective at eliminating pathogens from water. However, inadequate, interrupted, or intermittent treatment has repeatedly been associated with waterborne disease outbreaks. Contamination is not evenly distributed but rather affected by the number of pathogens in the source water, the age of the distribution system, the quality of the delivered water, and climatic events that can tax treatment plant operations. Private water supplies are not regulated by the USEPA and are generally not treated or monitored, although very few of the municipal systems involved in documented outbreaks exceeded the USEPA's total coliform standard in the preceding 12 mon (Craun et al. 2002).

We provide here estimates of waterborne infection and illness risks in the U.S. based on the total number of water systems, source water type, and total populations exposed. Furthermore, we evaluated all possible illnesses associated with the microbial infection and not just gastroenteritis. Our results indicate that 10.7 M infections/yr and 5.4 M illnesses/yr occur in populations served by community groundwater systems; 2.2 M infections/yr and 1.1 M illnesses/yr occur in noncommunity groundwater systems; and 26.0 M infections/yr and 13.0 M illnesses/yr occur in municipal surface water systems. The total estimated number of waterborne illnesses/yr in the U.S. is therefore estimated to be 19.5 M/yr. Others have recently estimated waterborne illness rates of 12 M cases/yr (Colford et al. 2006) and 16 M cases/yr (Messner et al. 2006), yet our estimate considers all health outcomes associated with exposure to pathogens in drinking water rather than only gastrointestinal illness.

Drinking water outbreaks exemplify known breaches in municipal water treatment and distribution processes and the failure of regulatory requirements to ensure water that is free of human pathogens. Water purification technologies applied at the point-of-use (POU) can be effective for limiting the effects of source water contamination, treatment plant inadequacies, minor intrusions in the distribution system, or deliberate posttreatment acts (i.e., bioterrorism). Epidemiological studies are conflicting on the benefits of POU water treatment. One prospective intervention study found that consumers of reverse-osmosis (POU) filtered water had 20%–35% less gastrointestinal illnesses than those consuming regular tap water, with an excess of 14% of illness due to contaminants introduced in the distribution system (Payment 1991, 1997). Two other studies using randomized, blinded, controlled trials determined that the risks were equal among groups supplied with POU-treated water compared to untreated tap water (Hellard et al. 2001; Colford et al. 2003). For immunocompromised populations, POU water treatment devices are recommended by the CDC and USEPA as one treatment option for reducing risks of *Cryptosporidium* and other types of infectious agents transmitted by drinking water. Other populations, including those experiencing "normal" life stages such as pregnancy, or those very young or very old, might also benefit from the utilization of additional water treatment options beyond the current multibarrier approach of municipal water treatment.

References

Abbaszadegan M, LeChevallier M, Gerba C (2003) Occurrence of viruses in US groundwaters. J Am Water Works Assoc 95:107–120.

Aboytes R, DiGiovanni G, Abrams FA, Rheinecker C, McElroy W, Shaw N, Lechevallier MW (2004) Detection of infectious *Cryptosporidium* in filtered drinking water. J Am Water Works Assoc 96:88–98.

Addiss DG, Pond RS, Remshak M, Juranek DD, Stokes S, Davis JP (1996) Reduction of risk of watery diarrhea with point-of-use water filters during a massive outbreak of waterborne *Cryptosporidium* infection in Milwaukee, Wisconsin, 1993. Am J Trop Med 54:549–553.

Allen MJ, Edberg SC, Reasoner DJ (2004) Heterotrophic plate count bacteria—what is their significance in drinking water? Int J Food Microbiol 92:265–274.

American Pregnancy Association (2006) Available online at http://www.americanpregnancy.org/main/statistics.html.

AWWA (2001) Reinvesting in drinking water structure: dawn of the replacement era. American Water Works Association, Denver, CO.

AWWA (2003) Water Stats 2002 Distribution Survey CD-ROM. American Water Works Association, Denver, CO.

AWWA and EES (2002) Finished water storage facilities. Available online at www.epa.gov/safewater/tcr/pdf/storage.pdf.

AWWA and AWWARF (1992) Water industry database: utility profiles. American Water Works Association and American Water Works Association Research Foundation, Denver, CO.

AWWASC (2002) Deteriorating buried infrastructure management challenges and strategies. American Water Works Association Service Company, Inc. Available online at http://www.epa.gov/safewater/tcr/pdf/infrastructure.pdf.

Barbeau B, Payment P, Coallier J, Clement B, Prevost M (2000) Evaluating the risk of infection from the presence of *Giardia* and *Cryptosporidium* in drinking water. Quant Microbiol 2:37–54.

Baribeau H, Pozos NL, Boulos L, Crozes GF, Gagnon GA, Rutledge S, Skinner D, Hu Z, Hofmann R, Andrews RC, Wojcicka L, Alam Z, Chauret C, Andrews SA, Dumancis R, Warn E (2005) Impact of Distribution System Water Quality on Disinfection Efficacy. AWWARF, Denver, CO.

Barwick MS, Levy DA, Craun GF, Beach MJ, Calderon RL (2000) Surveillance for waterborne-disease outbreaks—United States, 1997–1998. CDC, MMWR Surveillance Summaries 49(SS04):1–35.

Bennett M (2006) Bats and human emerging diseases. Epidemiol Infect 134:905–907.

Blackburn RS, Craun GF, Yoder JS, Hill V, Calderon RL, Chen N, Lee SH, Levy DA, Beach MJ (2004) Surveillance for waterborne-disease outbreaks associated with drinking water—United States, 2001–2002. MMWR 53(SS-8):23–45.

Borchardt MA, Bertz PD, Spencer SK, Battigelli DA (2003a) Incidence of enteric viruses in groundwater from household wells in Wisconsin. Appl Environ Microbiol 69:1172–1180.

Borchardt MA, Chyou PH, DeVries EO, Belongia EA (2003b) Septic system density and infectious diarrhea in a defined population of children. Environ Health Perspect 111:742–748.

Borchardt MA, Haas NL, Hunt RJ (2004) Vulnerability of drinking-water wells in La Crosse, Wisconsin, to enteric-virus contamination from surface water contributions. Appl Environ Microbiol 70:5937–5946.

Calderon RG (1991) Bacteria colonizing point of use, granular activated carbon filters and their relationship to human health. Final report CR-813978-01-0. U.S. Environmental Protection Agency, Cincinnati, OH.

Calderon RL (2001) Microbes in drinking water: recent epidemiological research to assess waterborne risks. In: Craun GF, Hauchman FS, Robinson DE (eds) Microbial Pathogens and Disinfection By-Products in Drinking Water. Health Effects and Management of Risks. ILSI Press, Washington, DC, pp 137–147.

Calderon RL (2004) Measuring benefits of drinking water technology: ten years of drinking water epidemiology. NEWWA Water Quality Symposium, May 20, 2004, Boxborough, MA.

Calderon RL, Craun GF (2006) Estimates of endemic waterborne risks from community-intervention studies. J Water Health 4(suppl 2):89–99.

Camper AK, LeChevallier ML, Broadaway SC, McFeters GA (1985) Growth and persistence of pathogens on granular activated carbon. Appl Environ Microbiol 50:1178–1382.

CDC (1993) Surveillance for waterborne-disease outbreaks—United States, 1991–1992. Centers for Disease Control and Prevention. MMWR 42(SS05):1–22.

CDC (2001) *Helicobacter pylori*: fact sheet for health care providers. Centers for Disease Control and Prevention. Available online at www.cdc.gov/ulcer/files/hpfacts.pdf.

CDC (2005) National Estimates on Diabetes. National Center for Chronic Disease Prevention and Health Promotion, Centers for Disease Control and Prevention. http://www.cdc.gov/diabetes/pubs/estimates.htm#prev.

CGER (2001) Under the weather: climate, ecosystems, and infectious disease. Commission on Geosciences, Environment and Resources. National Academies Press, Washington, DC.

Chen X, Keithly JS, Paya CV, LaRusso NF (2002) Cryptosporidiosis. N Engl J Med 346:1723–1731.
Chaidez C, Gerba C (2004) Comparison of the microbiologic quality of point-of-use (POU)-treated water and tap water. Int J Environ Health Res 14:253–260.
Clark RM, Geldreich EE, Fox KR, Rice WW, Johnson CH, Goodrich JA, Barnick JA, Abdesaken F (1996) Tracking a *Salmonella* serovar typhimurium outbreak in Gideon, Missouri: Role of contamination propagation modeling. J Water Supply Res Tech Aqua 45:171–183.
Colford JM Jr, Rees JR, Wade TJ, Khalakdina A, Hilton JF, Ergas IJ, Burns S, Benker A, Ma C, Bowen C, Mills D, Vugia D, Juranek D, Levy D (2002) Participant blinding and gastrointestinal illness in a randomized, controlled trial of an in-home drinking water intervention. Emerg Infect Dis 8:29–36.
Colford JM Jr, Wade TJ, Sandhu SK, Wright CC, Lee S, Shaw S, Fox K, Burns S, Benker A, Brookhart MA, Van Der Laan MJ, Levy DA (2005) A randomized controlled trial of in-home drinking water intervention to reduce gastrointestinal illness. Am J Epidemiol 161:472–482.
Colford JM, Roy S, Beach MJ, Hightower A, Shaw SE, Wade TJ (2006) A review of household drinking water trials and an approach to the estimation of endemic waterborne gastroenteritis in the United States. J Water Health 4(suppl 2):71–88.
Covello VT, Merkhofer MW (1993) Risk Assessment Methods. Plenum Press, New York.
Cox P, Griffith M, Angeles M, Deere D, Ferguson C (2005) Concentration of pathogens and indicators in animal feces in the Sydney watershed. Appl Environ Microbiol 71:5929–5934.
Craun GF, Calderon RL (2001) Waterborne disease outbreaks caused by distribution system deficiencies. J Am Water Works Assoc 93:64–75.
Craun GF, Nwachuku N, Calderon RL, Craun MF (2002) Outbreaks in drinking-water systems, 1991–1998. J Environ Health 65:16–23.
Craun MF, Craun GF, Calderon RL, Beach MJ (2006) Waterborne outbreaks in the United States. J Water Health 4(suppl 2):19–30.
Curriero FC, Patz JA, Rose JB, Lele S (2001) The association between extreme precipitation and waterborne disease outbreaks in the United States, 1948–1994. Am J Public Health 91:1194–1199.
Deb AK, Hasit YJ, Grablutz FM (1995) Distribution System Performance Evaluation. AWWARF, Denver, CO.
Edberg SC, Allen MJ (2004) Virulence and risk from drinking water of heterotrophic plate count bacteria in human population groups. Int J Food Microbiol 92:255–263.
Edwards M, Marshall B, Zhang Y, Lee Y (2005) Unintended consequences of chloramine hit home. In: Proceedings of the WEF Disinfection Conference, Mesa, AZ.
Egorov A, Ford T, Tereschenko A, Drizhd N, Segedevich I, Fourman V (2002) Deterioration of drinking water quality in the distribution system and gastrointestinal morbidity in a Russian city. Int J Environ Health Res 12:221–233.
Elitsur Y, Short JP, Neace C (1998) Prevalence of *Helicobacter pylori* infection in children from urban and rural West Virginia. Dig Dis Sci 43:773–778.
Fallacara DM, Monahan CM, Morishita TY, Bremen CA, Back RF (2004) Survey of parasites and bacterial pathogens from free-living waterfowl in zoological settings. Avian Dis 48:759–767.

Flannery B, Gelling LB, Vugia DJ, Weintraub JM, Salerno JJ, Conroy MJ, Stevens VA, Rose CE, Moore MR, Fields BS, Besser RE (2006) Reducing *Legionella* colonization of water systems with monochloramine. Emerg Infect Dis. Available at http://www.cdc.gov/ncidod/EID/vol12no04/05-1101.htm.

Fout GS, Martinson BC, Moyer MW, Dahling DR (2003) A multiplex reverse transcription-PCR method for detection of human enteric viruses in groundwater. Appl Environ Microbiol 69:3158–3164.

Frost FJ, Kunde TR, Harder L, Muller TI (2006) Preliminary Report—Northwest Epidemiological Enteric Disease Study. Unpublished report. Lovelace Clinic Foundation, Albuquerque, NM.

Gadgil AJ (1998) Drinking water in developing countries. Annu Rev Energy Environ 23:253–286.

Gerba CP (2003) Presentation at the National Science Foundation/Water Quality Center Bi-Annual meeting, December 2002, Tempe, AZ.

Gerba CP, Rose JB, Haas CN (1996a) Sensitive populations: who is at the greatest risk? Int J Food Microbiol 30:113–123.

Gerba CP, Rose JB, Haas CN, Crabtree CD (1996b) Waterborne rotavirus: a risk assessment. Water Res 30:2929–2940.

Glynn M, Rhodes P (2005) Estimated HIV prevalence in the United States at the end of 2003. National HIV Prevention Conference, June 2005, Atlanta, GA.

Goh S, Reacher M, Casemore DP, Verlander NQ, Charlett A, Chalmers RM, Knowles M, Pennington A, Williams J, Osborn K, Richards S (2005) Sporadic cryptosporidiosis decline after membrane filtration of public water supplies, England, 1996–2002. Emerg Infect Dis 11:251–259.

Grigg NS (2005) Assessment and renewal of water distribution systems. J Am Water Works Assoc 97:58–68.

Haas C (1999) Benefits of using a disinfectant residual. J Am Water Works Assoc 90:65–67.

Haas CN, Trussell RR (1998) Framework for assessing reliability of multiple, independent barriers in potable water reuse. Water Sci Technol 38:1–8.

Haas CN, Rose JB, Gerba CP, Regli S (1993) Risk assessment of virus in drinking water. Risk Anal 13:545–552.

Haas CN, Chitluru RB, Gupta M, Pipes WO, Burlingame GA (1998) Development of disinfection guidelines for the installation and replacement of water mains. AWWARF, Denver, CO.

Hancock CM, Rose JB, Callahan M (1998) *Cryptosporidium* and *Giardia* in U.S. groundwater. J Am Water Works Assoc 90(3):58–61.

Hawkins M, DeLong S, Marcus R, Jones T, Shallow S, Morse D, McCombs K, Courtney A, Medus C, Shiferaw B, Imhoff B, the EIP FoodNet Working Group (2002) The burden of diarrheal illness in FoodNet, 2000–2001 (abstract). In: Program and Abstracts of the International Conference on Emerging Infectious Diseases, Atlanta, GA. March 2002. American Society for Microbiology, Alexandria, VA, p 1. Available online at http://www.cdc.gov/foodnet/publications/2002/hawkins_2002.pdf.

Heirholzer JC (1992) Adenoviruses in the immunocompromised host. Clin Microbiol Rev 5:262–274.

Hellard ME, Sinclair MI, Forbes AB, Fairley CK (2001) A randomized, blinded, controlled trial investigating the gastrointestinal health effects of drinking water quality. Environ Health Perspect 109:773–778. Available online, 1 August 2001, at http://ehpnet1.niehs.nih.gov/docs/2001/109p773-778hellard/abstract.html.

Hellard ME, Sinclair MI, Dharmage SC, Bailey JF, Fairley CK (2002) The rate of gastroenteritis in a large city before and after chlorination. Int J Environ Health Res 12:355–360.

Herbarth O, Krumbiegel P, Fritz GJ, Richter M, Schlink U, Müller DM, Richter T (2001) *Helicobacter pylori* prevalences and risk factors among school beginners in a German urban center and its rural county. Environ Health Perspect 109: 573–577.

Herrick D (1997) Cross-connections and backflow. Water Well J 51:67–70.

Imhoff B, Morse D, Shiferaw B, Hawkins, M, Vugia D, Lance-Parker S, Hadler J, Medus C, Kennedy M, Moore MR, Van Gilder T, Group FW (2004) Burden of self-reported acute diarrheal illness in FoodNet surveillance areas, 1998–1999. Clin Infect Dis 38(suppl 3):S219–S226.

James KE, Block DA, Lee KK, Kraemer HC, Fuller RK (1996) An index for assessing blindness in a multi-centre clinical trial: disulfiram for alcohol cessation: a VA cooperative study. Stat Med 15:1421–1434.

Jemal A, Murray T, Ward E, Samuels A, Tiwari RC, Ghafoor A, Feuer EJ, Thun MJ (2005) Cancer statistics, 2005. CA Cancer J Clin 55:10–30.

Jones TF, McMillian MB, Scallan E, Frenzen PD, Cronquist AB, Thomas S, Angulo, FJ (2007) A Population-Based Estimate of the Substantial Burden of Diarrheal Disease in the United States; FoodNet, 1996–2003. Centers for Disease Control and Prevention, Atlanta, GA 135:293–301.

Karim M, Abbaszadegan M, LeChevallier MW (2003) Potential for pathogen intrusion during pressure transients. J Am Water Works Assoc 95:134–146.

Kirmeyer G, Richards W, Smith CD (1994) An assessment of water distribution systems and associated research needs. AWWARF, Denver, CO.

Kirmeyer GK, Freidman M, Martel K, Howie D, LeChevallier M, Abbaszadegan M, Karim M, Funk J, Harbour J (2001) Pathogen intrusion into the distribution system. AWWARF, Denver, CO.

Korich DG, Mead JR, Madore MS, Sinclair NA, Sterling CR (1990) Effects of ozone, chlorine dioxide, chlorine, and chloramine on *Cryptosporidium parvum* oocyst viability. Appl Environ Microbiol 56:1423–1428.

Kramer MH, Herwaldt BL, Calderon RL, Juranek DD (1996) Surveillance for waterborne-disease outbreaks—United States, 1993–1994. CDC MMWR Surveillance Summaries 45(SS01):1–33.

Kunde TR, Frost FJ, Nelson LS, Harter L, Craun MF, Craun GF, Calderon RL (2006) Estimates of endemic waterborne risks. Community intervention study: Reduced gastrointestinal illness rates associated with improved water treatment. Preliminary report: A 98. Lovelace Clinic Foundation, Albuquerque, NM.

LeChevallier MW, Gullick RW, Karim MR, Friedman M, Funk JE (2003a) The potential for health risks from intrusion of contaminants into the distribution system from pressure transients. J Water Health 1:3–14.

LeChevallier MW, Gullick RW, Karim M (2003b) The potential for health risks from intrusion of contaminants into the distribution system from pressure transients. Distribution system white paper. http://www.epa.gov/safewater/tcr/pdf/intrusion.pdf.

Lee EJ, Schwab KJ (2005) Deficiencies in drinking water distribution systems in developing countries. J Water Health 3:109–127.

Lee SH, Levy DA, Craun GF, Beach MJ, Calderon RL (2002) Surveillance for waterborne-disease outbreaks—United States, 1999–2000. CDC MMWR Surveillance Summaries 51(SS08):1–28.

Lee JJ, Schwartz P, Sylvester P, Crane L, Haw J, Chang H, Kwon HJ (2003) Impacts of Cross-Connections in North American Water Supplies. AWWARF, Denver, CO.

Levy DA, Bens MS, Craun GF, Calderon RL, Herwaldt BL (1998) Surveillance for waterborne-disease outbreaks—United States, 1995–1996. CDC MMWR Surveillance Summaries 47(SS05):1–34.

Lew JF, Glass RI, Gangarosa RE, Cohen IP, Bern C, Moe CL (1991) Diarrheal deaths in the United States, 1979–1987. A special problem for the elderly. JAMA 265:3280–3284.

Liang JL, Dziuban EJ, Craun GF, Hill V, Moore MR, Gelting RJ, Calderon RL, Beach MJ, Roy SL (2006) Surveillance for waterborne disease and outbreaks associated with drinking water and water not intended for drinking—United States, 2003–2004. MMWR 55(12):31–65.

Lindsay JA (1997) Chronic sequelae of food-borne disease. Emerg Infec Dis 3:443–452.

Mansfield LS, Gajadhar AA (2004) *Cyclospora cayetanensis*, a food- and waterborne coccidian parasite. Vet Parasitol 126:73–90.

McConnell S, Horrocks M, Sinclair M, Fairley CK (2001) Changes in the incidence of gastroenteritis and the implementation of public water treatment. Int J Environ Health Res 11:299–308.

Medema GJ, Teunis PF, Havelaar AH, Haas CN (1996) Assessment of the dose–response relationship of *Campylobacter jejuni*. Int J Food Microbiol 30:101–111.

Messner MJ, Chappell CL, Okhuysen PC (2001) Risk assessment for *Cryptosporidium*: a hierarchical Bayesian analysis of human dose response data. Water Res 35:3934–3940.

Messner M, Shaw S, Regli S, Rotert K, Blank V, Soller J (2006) An approach for developing a national estimate of waterborne disease due to drinking water and a national estimate model application. J Water Health 4(suppl 2):201–240.

Morris RD, Levin R (1995) Estimating the incidence of waterborne infectious disease related to drinking water in the United States. In: Reichard E, Zapponi G (eds) Assessing and managing health risks from drinking water contamination: approaches and applications. Proceedings of a symposium held in Rome, September 1994. Publication 233. International Association of Hydrological Sciences, Wallingford, UK.

NRC (2006) Drinking water distribution systems: assessing and reducing risks. National Research Council. The National Academies Press, Washington, DC, pp 90–137.

Nwachuku N, Gerba CP (2006) Health risks of enteric viral infections in children. Rev Environ Contam Toxicol 186:1–56.

Nwachuku N, Gerba CP, Oswald A, Mashadi FD (2005) Comparative inactivation of adenovirus serotypes by UV light disinfection. Appl Environ Microbiol 71: 5633–5636.

Park SR, Mackay WG, Reid DC (2001) *Helicobacter sp.* recovered from drinking water biofilm sampled from a water distribution system. Water Res 35:1624–1626.

Parkin RT (2000) Issues in modeling rare health events such as chronic sequelae associated with microbial pathogens. Annual Meeting of the Society for Risk Analysis. Center for Risk Science and Public Health, Arlington, VA, December 2000.

Payment P (1999) Poor efficacy of residual chlorine disinfectant in drinking water to inactivate waterborne pathogens in distribution systems. Can J Microbiol 45: 709–715.

Payment P, Richardson L, Siemiatycki J, Dewar R, Edwardes M, Franco E (1991) A randomized trial to evaluate the risk of gastrointestinal disease due to consumption of drinking water meeting current microbiological standards. Am J Public Health 81:703–708.

Payment P, Siemiatycki J, Richardson L, Renaud G, Franco E, Prevost M (1997) A prospective epidemiological study of gastrointestinal health effects due to the consumption of drinking water. Int J Environ Health Res 7:5–31.

Pepper IL, Rusin P, Quintanar DR, Haney C, Josephson KL, Gerba CP (2004) Tracking the concentration of heterotrophic plate count bacteria from the source to the consumer's tap. Int J Food Microbiol 9:289–295.

Pierson G, Martel K, Hill A, Burlingame G, Godfree A (2001) Methods to prevent microbiological contamination associated with main rehabilitation and replacement. AWWARF, Denver, CO.

Propato M, Uber JG (2004) Vulnerability of water distribution systems to pathogen intrusion: how effective is a disinfectant residual? Environ Sci Technol 38: 3713–3722.

Pryor M, Springthorpe S, Riffard S, Brooks T, Huo Y, Davis, G Satter SA (2004) Investigation of opportunistic pathogens in municipal drinking water under different supply and treatment regimes. Water Sci Technol 50:83–90.

Raina PS, Pollari FL, Teare GF, Goss MJ, Barry DA, Wilson JB (1999) The relationship between *E. coli* indicator bacteria in well-water and gastrointestinal illnesses in rural families. Can J Public Health 90:172–175.

Regli S, Rose JB, Haas CN, Gerba CP (1991) Modeling the risk from *Giardia* and viruses in drinking water. J Am Water Works Assoc 19:76–84.

Regli SE, Berger PS, Grubbs TR (2003) Control of drinking water pathogens and disinfection byproducts. In: Pontius FW (ed) Drinking Water Regulation and Health. Wiley, New York, pp 277–305.

Rennecker JL, Marinas BJ, Owens JH, Rice EW (1999) Inactivation of *Cryptosporidium parvum* (oo)cysts with ozone. Water Res 33:2481–2488.

Rollinger Y, Dott W (1987) Survival of selected bacterial species in sterilized activated carbon filters and biological activated filters. Appl Environ Microbiol 53:777–781.

Smith HV, Grimason AM (2003) *Giardia* and *Cryptosporidium* in water and wastewater. In: The Handbook of Water and Wastewater Microbiology. Academic Press, London, pp 698–756.

Snead MC, Olivieri VP, Kawata K, Kruse CW (1980) The effectiveness of chlorine residuals in inactivation of bacteria and viruses introduced by post-treatment contamination. Water Res 14:403–408.

Stine SW, Pepper IL, Gerba CP (2005) Contribution of drinking water to the weekly intake of heterotrophic bacteria from diet in the United States. Water Res 39: 257–263.

Thomas KM, Charron DF, Waltner-Toews D, Schuster C, Maarouf AR, Holt JD (2006) A role of high impact weather events in waterborne disease outbreaks in Canada, 1975–2001. Int J Environ Health Res 16:167–180.

Thurston-Enriquez JA, Haas CN, Jacangelo J, Riley K, Gerba CP (2003) Inactivation of feline calicivirus and adenovirus type 40 by UV radiation. Appl Environ Microbiol 69:577–582.

Tobin-D'Angelo MJ, Blass MA, del Rio C, Halvosa JS, Blumberg HM, Horsburgh CR (2004) Hospital water as a source of complex isolates in respiratory specimens. J Infect Dis 189:98–104.
Trussell RR (1999) Safeguarding distribution system integrity. J Am Water Works Assoc 91:46–54.
US Census Bureau (2005) American Community Survey. Available at http://factfinder.census.gov.
US Department of Health and Human Services (2005) Annual Report of the U.S. Organ Procurement and Transplantation Network and the Scientific Registry of Transplant Recipients: Transplant Data 1995–2004. Health Resources and Services Administration, Healthcare Systems Bureau, Division of Transplantation, Rockville, MD. Available at http://www.hrsa.gov/.
USEPA (2001) Drinking Water Infrastructure Needs Survey. EPA 816-F-01-001. Office of Water, U.S. Environmental Protection Agency, Washington, DC.
USEPA (2002) Potential contamination due to cross-connections and backflow and the associated health risks: An issues paper. EPA's Office of Ground Water and Drinking Water. Cross-Connection Control. August 13, 2002. Available at http://www.m_2.us/files/ccrwhite.pdf.
USEPA (2004) USEPA claims to meet drinking water goals despite persistent data quality shortcomings. Report no. 2004-P-0008. Available online at www.epa.gov/oig/reports/2004/20040305-2004-P-0008.pdf.
USEPA (2006a) Prepublication of ground water rule federal register notice. national primary drinking water regulations: ground water rule. 40 CFR Parts 9, 141, and 142. EPA-HQ-OW-2002-0061; FRL-RIN-2040- AA97. U.S. Environmental Protection Agency, Washington, DC.
USEPA (2006b) Public Drinking Water Systems: Facts and Figures. Available at http://www.epa.gov/safewater/data/getdata.html.
Vacrewijck MJM, Huys G, Palomino JC, Swings J, Portaels F (2005) Mycobacteria in drinking water distribution systems: ecology and significance for human health. FEMS Microbiol Rev 29:911–934.
Velkoff VA, DeBarros KA (2005) 65+ in the United States: 2005. Current population reports. U.S. Census Bureau, U.S. Government Printing Office, Washington, DC, pp 23–209.
WQA (Water Quality Association) (2001) The 2001 national consumer water quality survey. Water Quality Association, Lisle, IL. Available online at www.wqa.org.
WQA (Water Quality Association) (2002) Heterotrophic bacteria in drinking water from POU & POE devices. Water Quality Association, Lisle, IL. Available at http://www.wqa.org/pdf/technical/HPC200%5B5%5D.pdf.
Westrell T, Bergstedt O, Stenstrom TA, Ashbolt NJ (2003) A theoretical approach to assess microbial risks due to failures in drinking water systems. Int J Environ Health Res 13:181–197.
WHO/NSF (2003) HPC and drinking-water safety: the significance of heterotrophic plate counts for water quality and human health. Available at http://www.nsf.org/conference/hpc/hpc_proceedings.html.

Manuscript received January 9; accepted January 12, 2007.

Rev Environ Contam Toxicol 192:159–195

Biological Removal of Nitrogen from Wastewater

Guibing Zhu, Yongzhen Peng, Baikun Li, Jianhua Guo, Qing Yang, and Shuying Wang

Contents

I. Introduction

The removal of ammonia from wastewater has become a worldwide emerging concern because ammonia is toxic to aquatic species and causes

---

Communicated by Dr. David M. Whitacre

G. Zhu, Y. Peng (✉), J. Guo
School of Municipal and Environmental Engineering, Harbin Institute of Technology, Harbin 150090, P.R. China

B. Li
Department of Civil and Environmental Engineering, University of Connecticut, 261 Glenbrook Rd., UNIT-2037, Storrs, CT 06269

Y. Peng, Q. Yang, S. Wang
Key Laboratory of Beijing Water Quality Science and Water Environment Recovery Engineering, Beijing University of Technology, Beijing 100022, P.R. China

G. Zhu
SKLEAC, Research Center for Eco-Environmental Sciences, Chinese Academy of Sciences, Beijing 100085, P.R. China

eutrophication in natural water environments (Tchobanoglous et al. 2003). Nitrogen compounds in wastewater can only be effectively removed by biological approaches (EPA 1993; Zhu et al. 2007a,b). Based on the microbial nitrogen cycle and the metabolism of inorganic nitrogen compounds (Fig. 1), many biological technologies and processes have been developed and implemented for nitrogen removal from wastewater, such as predenitrification (Anoxic/Oxic), modified Bardenpho, Bio-denitro, sequencing batch reactor (SBR), oxidation ditch (OD), step feeding, anaerobic/anoxic/aerobic (A^2/O), and University of Cape Town (UCT) processes (Wentzel et al. 1992; Østgaard et al. 1997; Williams and Beresford 1998; Tchobanoglous et al. 2003; Pai et al. 2004). These processes have been widely employed in wastewater treatment plants for nitrification and denitrification (EPA 1993). However, with the effluent discharge standards having become more stringent (<10 mg total nitrogen/L), conventional processes cannot meet the new requirements (Khin and Annachhatre 2004).

Several novel nitrogen removal processes have been developed to enhance nitrification and denitrification. This review focuses on these novel processes, including simultaneous nitrification and denitrification (SND), shortcut nitrification and denitrification, anaerobic ammonium oxidation (ANAMMOX), aerobic deammonitrification, completely autotrophic nitrogen removal over nitrite (CANON), oxygen-limited autotrophic nitrification-denitrification (OLAND) processes (Muller et al. 1995; Strous et al. 1999; Fux et al. 2002; Third et al. 2001; Schmidt et al. 2003; Nielsen et al. 2005; Peng and Zhu 2006). Particularly, this review presents a critical comparison of various biological processes, discusses the key control parameters, and summarizes the current research status of functional microorganisms for nitrogen removal. Moreover, several challenging and unsolved problems of these processes are addressed.

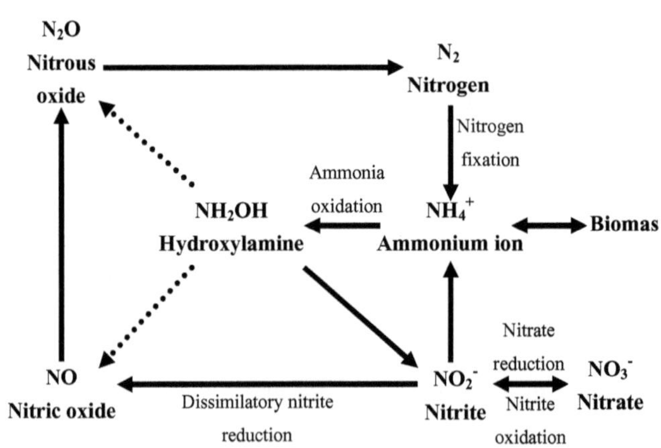

Fig. 1. Microbial nitrogen cycle. (From Rick and Stuart 2001.)

II. Conventional Biological Technologies for Nitrogen Removal

A. Mechanism and Principle

Conventional microbial nitrogen removal is based on autotrophic nitrification and heterotrophic denitrification. In the first step of nitrification, ammonia-oxidizing bacteria (AOB) oxidize ammonium (NH_4^+) to nitrite (NO_2^-) via hydroxylamine (NH_2OH) (reactions 1 and 2, below). Membrane-bound ammonia monooxygenase (AMO) and hydroxylamine oxidoreductase (HAO) are involved in these two reactions. In the second step, nitrite-oxidizing bacteria (NOB) oxidize nitrite to nitrate (NO_3^-) with the involvement of membrane-bound nitrite oxidoreductase (NOR) (reaction 3).

$$NH_3 + O_2 + 2[H] \xrightarrow{AMO} NH_2OH + H_2O \qquad (1)$$

$$HN_2OH + 0.5O_2 \xrightarrow{HAO} NHO_2 + 2H^+ + 2e^- \qquad (2)$$

$$NO_2^- + 0.5O_2 \xrightarrow{NOR} NO_3^- \qquad (3)$$

In anoxic denitrification, NO_3^- and NO_2^- are reduced to gaseous nitrogen with a variety of electron donors, such as methanol, acetate, and organic substances in wastewater (reactions 4 and 5).

$$2NO_3^- + 10H^+ + 10e^- \rightarrow N_2 + 2OH^- + 4H_2O \qquad (4)$$

$$2NO_2^- + 6H^+ + 6e^- \rightarrow N_2 + 2OH^- + 2H_2O \qquad (5)$$

B. Typical Processes for Biological Removal of Nitrogen

Many biological nitrogen removal processes have been developed, including Bardenpho, predenitrification, postdenitrification, SBR, OD, and step feeding (Tchobanoglous et al. 2003; Dapena-Mora et al. 2004; Khin and Annachhatre 2004; Zhu et al. 2005, 2007a,b). The advantages and limitations of these processes are summarized in Table 1.

In most countries, especially in China, about 80% of wastewater treatment plants use the predenitrification [i.e., anoxic/oxic (A/O)] process for biological nitrogen removal (Zhu 2006). Predenitrification has distinct advantages for nitrogen removal. With influent first entering the anoxic denitrification zone, organic carbon sources serve as electron donors for denitrification and are biodegraded by denitrifying bacteria. This method can improve nitrogen removal efficiency and shorten the aerobic duration. However, because of the configuration of A/O processes, the NO_X-N concentration in effluent equals that of internally recycled wastewater, which results in an overall low nitrogen removal efficiency (Baeza et al. 2004; Rosso and Stenstrom 2005). For instance, based on the theory

Table 1. Advantages and Limitations of Conventional Biological Nitrogen Removal Processes.

Process	Advantages	Limitations
Bardenpho (4-stage)	Total N concentration less than 3 mg/L possible	Large reactor volume is required Second anoxic tank has low efficiency
Predenitrification	Very adaptable to existing activated sludge processes 5–8 mg/L total N is achievable Reparation of alkalinity because of denitrification	N-removal capability is a function of internal cycle DO control is required before recycle
Postdenitrification	Capable of achieving total N levels less than 3 mg/L	Higher operating cost due to additional carbon dosage Carbon dosage control is required
Bio-denitro	Large reactor volume is resistant to shock load 5–8 mg/L total N is achievable	More complex to operate Two reactors are required so as to increase construction cost
Sequencing batch reactor	Process is flexible and easy to design Quiescent settling provides low effluent TSS concentration Mixed liquid/solids cannot be washed out by hydraulic surges Secondary settling tank is not required	More complex to operate Effluent quality depends on reliable decanting facility Be not suitable for large plants
Oxidation ditch	Highly reliable process Simple operation Capable of treating shock/toxic loads without affecting effluent quality Economical process for small plants Well-stabilized sludge; low biosolids production	Large structure, greater space requirement Low F/M bulking is possible Some modifications are proprietary and license fees may be required Plant capacity expansion is more difficult Nitrogen removal capability is related to skills of operating staff and control methods
Step feeding	Distributes load to provide more uniform oxygen demand To minimize high clarifier solids loading in peak wet weather flows Adaptable to existing activated sludge processes With internal recycle in last stage, total N less than 5 mg/L possible.	N-removal capability is a function of flow distribution Flow split is not measured or known accurately DO control is required before recycle Flow split control is required to optimize operation

TSS, total suspended solids; DO, dissolved oxygen.

of predenitrification process (Chiou and Ouyang 2001), a 100% sludge recycle and 200% internal recycle can achieve only 75% nitrogen removal efficiency.

C. Disadvantages and Limitations of Conventional Processes

Because nitrification and denitrification are carried out by different microorganisms under different conditions, they should be designed and operated in separate time sequences or spaces (Lee et al. 2001). Consequently, a long retention time or a large volume is required to accomplish complete nitrogen removal. Moreover, a high level of oxygen, set as 4.2 g O_2/g NH_4^+-N, is required for nitrification (Bruce and Perry 2001), and a sufficient organic carbon source [2.86 g chemical oxygen demand (COD)/g NO_3^--N) is necessary for denitrification (Gradly and Lim 1980). A high level of external carbon sources (methanol, acetate, etc.) is normally added in the denitrification process when treating wastewater with high nitrogen concentration or low C/N ratio (Tam et al. 1992), which increases the operational cost for conventional biological processes. The limitations of low removal efficiency, high oxygen requirement, long retention time, and an external carbon source are the driving forces for developing new low-cost biological treatment processes for complete nitrogen removal (Jetten et al. 2002).

III. Novel Biological Processes for Nitrogen Removal

A. Simultaneous Nitrification and Denitrification (SND)

Mechanism and Advantages

Simultaneous nitrification and denitrification (SND) means that nitrification and denitrification occur concurrently in the same reactor (Keller et al. 1997; Helmer and Kunst 1998). There are two mechanisms for SND: physical and biological (Robertson and Kuenen 1984; Baumann et al. 1996; Hibiya et al. 2003). The conventional physical mechanism is that SND occurs as the consequence of dissolved oxygen (DO) concentration gradients within activated sludge flocs or biofilms due to diffusional limitation (Fig. 2). The nitrifiers exist in aerobic regions with DO higher than 1–2 mg/L, whereas the denitrifiers stay alive in anoxic zones with DO less than 0.5 mg/L. The presence of oxygen concentration gradients in activated sludge flocs and biofilm has been verified by microelectrode measurements (Snidaro et al. 1997; de Beer et al. 1998; Satoh et al. 2003; Li and Bishop 2004; Holman and Wareham 2005) and ^{15}N tracer techniques (Wyffels et al. 2003).

The biological mechanism for SND is more complicated than the physical ones and is contradictory to the traditional "engineering" conception of autotrophic aerobic nitrification and heterotrophic anoxic denitrification. Several species of heterotrophic nitrifiers and aerobic denitrifiers have been identified in wastewater and night soil treatment systems (Patureau et al.

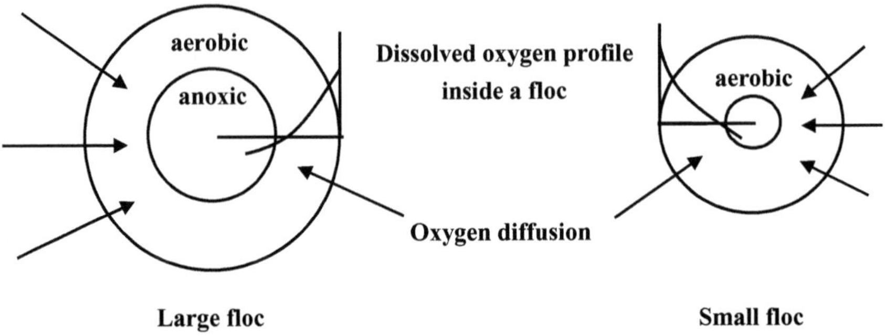

Fig. 2. Schematic of oxygen concentration profile within a microbial floc. Reprinted form Pochana and keller 1999, with permission from IWA Publishing.

1998; Hu and Kung 2000; Kim et al. 2005). *Alcaligenes* sp., *Corynebacterium* sp., *Acinetobacter* sp., *Xanthomonas* sp., and *Bacillus* strains were identified as heterotrophic nitrifiers (Castignetti and Gunner 1981; Castignetti and Hollocher 1982; Kshirsagar et al. 1995; Hu and Kung 2000; Kim et al. 2005). *Thiosphaera pantotropha* was identified as both heterotrophic nitrifier and aerobic denitrifier (Gupta 1997). *Paracoccus denitrificans,* an aerobic denitrifier isolated from activated sludge (Robertson et al. 1988, 1995; Baumann et al. 1996), reduced nitrate even under oxygen saturation. Other aerobic denitrifier strains, such as *Microvirgula aerodenitrificans* (Patureau et al. 1998) and *Thaurea mechernichensis* (Scholten et al. 1999), have been isolated. From a microbiological point of view, SND has been regarded as the consequence of the oxidization of ammonia by heterotrophic nitrifiers and the reduction of nitrate or nitrite by aerobic denitrifiers (Robertson et al. 1988, 1995; Wyffels et al. 2003).

SND has significant advantages over conventional processes (Pochana et al. 1999; Zhang et al. 2005). With denitrification taking place concurrently with nitrification in aeration tanks the SND process can save the costs for anoxic tanks, and simplify the overall process design. SND is of particular interest when treating wastewaters with a low C:N ratio (<5), because the cost of an extra carbon source will be saved (Guo et al. 2005).

Key Control Factors

On-line monitoring DO and redox potential (ORP) is an efficient approach in the SND process (Zhao et al. 1999; Fuerhacker et al. 2000). SND occurs at a DO of 0.5 mg/L, under which condition the nitrification and denitrification rates are the same (Munch et al. 1996; Peng et al. 2001). However, SBR occurs at a wide ORP range, from –60 to –198 mV (Hanaki et al. 1990; Fuerhacker et al. 2000; Hu et al. 2005).

Other operational parameters, such as sludge retention time (SRT), hydraulic retention time (HRT), and pH, have significant influence on the SND process. Because heterotrophic nitrifiers grow more rapidly and have stronger tolerance to acidity than autotrophic nitrifiers, short SRT and an acidic environment would be favorable for their growth (Focht and Verstraete 1977; Killham 1986; van Niel 1991). A bench-scale SBR system achieved a complete removal of NH_4^+-N and COD with no NO_2^--N in the effluent at a C/N ratio of 11.1, SRT of 20d, and HRT of 1d. However, the nitrogen removal efficiency decreased gradually with increasing ammonium-loading rates and F/Ms (Chiu et al. 2007). Until now, the SND processes have been tested with consistent SRT and HRT. There is no research on the influence of SRT and HRT on the efficiency of SND. In addition, pH and free ammonia (FA) should be studied as critical parameters for SND via NO_2^- as they have significant effects on the competition between AOB and NOB.

Research Status and Unsolved Concerns

There are three major factors for the SND processes: carbon source, DO concentration, and floc size (Lee et al. 2001; Holman and Wareham 2005).

Organic carbon is critical for SND, because high biological oxygen demand (BOD) concentration causes the inhibition of autotrophic nitrifiers whereas low BOD leads to the deficiency of electron donors for denitrifiers (Tam et al. 1992). SND occurs well when treating municipal wastewater at a BOD of 100–150mg/L (Castignetti and Gunner 1981; Kim et al. 2005).

DO concentration is also important for SND (Pochana et al. 1999): it has a twofold effect on SND performance (Pochana and Keller 1999; Hu et al. 2005; Zhang et al. 2005). Low DO concentration suppresses nitrification while high DO concentration inhibits denitrification. Nitrification and denitrification rates became the same at a DO concentration of 0.5mg/L and achieved a complete SND (Munch et al. 1996). Zhao et al. (1999) found that an extended aeration duration in an intermittent aeration (IA) process favored sequential nitrification and denitrification (SQND). The optimal DO concentration for SND via nitrite was around 2.0 ± 2.5mg/L at the end of the aeration period in the IA process (Yoo et al. 1999).

Some researchers attribute the occurrence of SND to the size of activated sludge floc, which is normally 80–100μm (Li and Ganczarczyk 1990, 1993; Pochana and Keller 1999). SND is more likely to occur in the large-size floc (>125μm) because of the oxygen diffusion limitation, but the occurrence of SND in activated sludge flocs smaller than 20μm is unclear. If SND is detected in small floc sludge (Wilen and Balmer 1999), the current physical explanation of SND processes will be put in question.

B. Shortcut Nitrification and Denitrification

Mechanism and Advantages

Shortcut nitrification and denitrification, namely partial nitrification-denitrification, is the process in which nitrification and denitrification are correlated by NO_2^- instead of NO_3^-. As an intermediate product, NO_2^- is produced in nitrification and reduced to N_2 in the following NO_2^- denitrification (Fdz-Polanco et al. 1996; van Dongen et al. 2001; Peng et al. 2006). Compared with traditional nitrification and denitrification via NO_3^-, shortcut nitrification and denitrification has the following advantages (Beccari et al. 1983; Turk and Mavinic 1989; Peng and Zhu 2006):

1. 25% lower oxygen consumption in the aerobic phase implies 60% energy saving in the entire process.
2. The requirement for electron donors is as much as 40% lower in the anoxic phase.
3. NO_2^- denitrification rate is 1.5 to 2 times higher than NO_3^- denitrification rate.

Partial nitrification via NO_2^- is reported to be technically feasible and economically favorable, especially when treating wastewater with high ammonia concentration or low C:N ratio (Turk and Mavinic 1989; Villaverde et al. 1997).

The Single reactor system for High Ammonia Removal Over Nitrite (SHARON) process, the first full-scale process with NO_2^- as the intermediate product, is a cost-effective treatment system for total nitrogen removal from wastewater with high nitrogen concentrations (>550 mg/L). The system has been used for treating wastewater generated from dewatered primary sludge, waste-activated sludge, sludge dryers, and incinerators (van Dongen et al. 2001).

Key Control Factors

The inhibition of nitrite-oxidizing bacteria (NOB) is critical for shortcut nitrification and denitrification because NOB oxidize NO_2^- to NO_3^- and convert partial nitrification to complete nitrification (Picioreanu et al. 1997; Hellinga et al. 1998; Hidaka et al. 2002; Peng and Zhu 2006). Several parameters, including DO concentration, temperature, SRT, substrate concentration, aeration pattern, and chemical inhibitor, have been found to selectively inhibit NOB.

Dissolved Oxygen Concentration. Compared with ammonia oxidizing bacteria (AOB), NOB require a high DO concentration. The DO half-saturation value for oxygen ($K_{s,O}$), representing the affinity for oxygen, is 62 μM for NOB whereas it is 16 μM for AOB (Picioreanu et al. 1997; Schramm et al. 1999, 2000). Therefore, AOB dominate NOB at low DO concentration, which results in the accumulation of NO_2^- and the occurrence of partial nitrification and denitrification via NO_2^-.

Although a low DO concentration (<1.5 mg/L) is favorable for partial nitrification, it reduces nitrification rates, lowers COD removal efficiencies, and causes sludge bulking. Different DO concentrations have been reported for partial nitrification, ranging from 0.3 to 2.5 mg/L (Wyffels et al. 2004a,b). High DO concentrations (>2 mg/L) could convert partial nitrification to complete nitrification, whereas low concentrations (<0.5 mg/L) could reduce nitrification rat. A DO concentration of 1.0–1.5 mg/L has been found suitable for shortcut nitrification and denitrification in real municipal wastewater treatment, which has been verified (Hanaki et al. 1990; Hao et al. 2002a; Peng et al. 2003).

Temperature. Correlation of the maximum growth rate of nitrifying bacteria and temperature is described in the Arrhenius equation at temperatures of 5°–40°C (Anthonisen 1976):

$$\mu_{mt} = \mu_{m20} \exp\left[-\frac{E_a(20-t)}{293R(273+t)}\right] \tag{6}$$

in which μ_{mt} is the maximal specific growth rate (d^{-1}), μ_{m20} is the maximal specific growth rate at 20°C (d^{-1}), E_a is the activation energy (kJ/mol), and R is a constant of 8.314 (J/mol K).

Growth rates of AOB and NOB vary with temperature. AOB have a higher maximal specific growth rate (0.801 d^{-1}) than NOB (0.788 d^{-1}) at 20°C (Hellinga et al. 1998), while the specific growth rate of AOB (0.523 d^{-1}) was lower than that of NOB (0.642 d^{-1}) at 15°C. Therefore, NOB dominate AOB at temperatures below 15°C, and AOB outcompete NOB at temperatures above 20°C (Brouwer et al. 1996) (Fig. 3). A higher temperature not only promotes the growth of AOB but can also expand the growth rate differences between AOB and NOB (Balmelle 1992; Hunik 1993; Yoo et al. 1999).

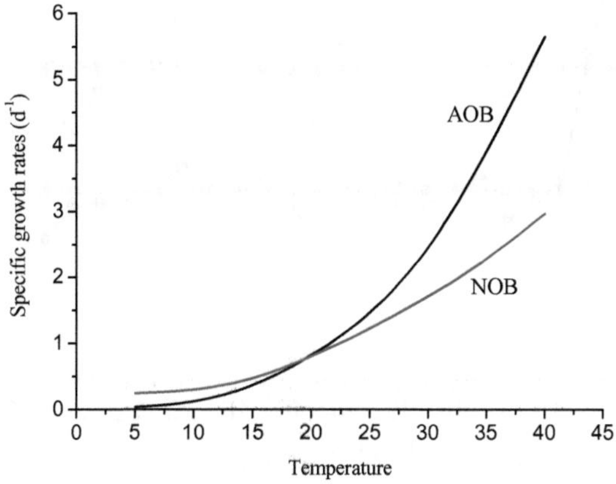

Fig. 3. Effect of temperature on growth rate of ammonia oxidizers and nitrite oxidizers.

The SHARON process has been successfully operated at 35°C, where AOB became dominant (Mulder et al. 2001). As a result of different bacterial growth rates in SHARON, a selection of microbial community should be made wherein NOB are washed out of the system while AOB are still retained along with denitrifying bacteria in the system. This operational mode allows a 25% reduction in oxygen consumption for nitrification and a 40% reduction in the external carbon source addition.

Sludge Age. AOB (e.g., *Nitrosomonas*) need a longer retention time than NOB (e.g., *Nitrobacter*) at temperatures below 15°C, while the trend was reversed at temperatures above 25°C (see Fig. 3). Thus, AOB and NOB can be selectively accumulated by appropriately adjusting SRT in a suspended-growth system (Hellinga et al. 1998).

SRT is equal to HRT in SHARON. Nitrogen was removed via nitrite in a SHARON process with an oxic HRT below 2 d. At an oxic HRT of approximately 1.5 days, the COD/N ratio clearly illustrates the metabolic pathways from ammonia to nitrogen via nitrite (Fig. 4).

Aeration Pattern. The aeration pattern has been proposed as an alternative to SRT for partial nitrification control (Hidaka et al. 2002). Aeration duration is inversely related to the extent of partial nitrification, because partial nitrification will be converted to complete nitrification at long

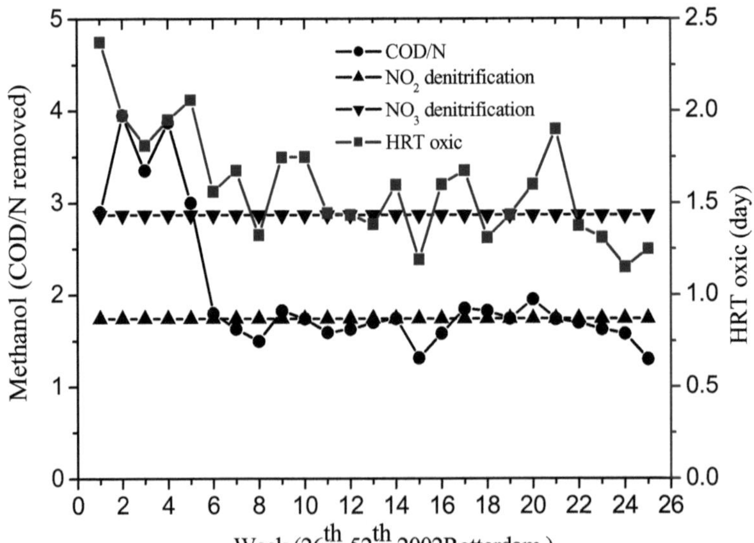

Fig. 4 Continuous operation chemical oxygen demand (COD)/N-removal nitrite pathway in Rotterdam *SHARON* system.

aeration periods (Turk and Mavinic 1989). Turk and Mavinic (1987) observed NO_2^- accumulated during a transition from anoxic to aerobic condition. This accumulation persisted 2–3 hr in the aerobic condition. Intermittent aeration favors partial nitrification (Yoo et al. 1999; Pollice et al. 2002). Peng et al. (2004a) reported that partial nitrification was successfully completed using the aeration control strategy, even though the temperature decreased from 32°C to 21°C.

Substrate Concentration and Load. AOB are divided into two groups according to cell growth rates: slow-growing and fast-growing (Zheng et al. 2004). Slow-growth bacteria, referred to as K strategists, have high affinity to substrate and are dominant at low substrate concentrations, whereas fast-growth bacteria, referred to as R strategists, have low affinity to substrate and thrive at high substrate concentrations. Because ammonia concentrations are normally below 5 mg/L in wastewater treatment processes to meet the discharge requirements, K strategists may be dominant. R strategists were found to become dominant in partial nitrification processes at high ammonia concentrations (>50 mg/L) (Surmacz-Gorska et al. 1997).

Inhibitors. Several inhibitors suppress NOB and lead to partial nitrification. Ag, Hg, Ni, Cr, Zn, Cu, and Pb, listed in increasing order of toxicity, inhibit nitrification (Camilla et al. 1998). Organic compounds such as aniline, ortho-cresol, and phenol exhibit stronger inhibitions on NOB than on AOB. Wastewater with these compounds might inhibit NOB and cause the accumulation of nitrite (Neufeld et al. 1986). Oxidants such as ClO_2^- and chlorate also inhibit NOB (Belser and Mays 1980). Seawater or saline wastewater containing a high level of ClO_2^- can achieve shortcut nitrification (Peng et al. 2004c).

High concentrations of free nitrous acid (HNO_2) and free ammonia (FA) also have adverse impacts on nitrification (Wouter et al. 1999; Villaverde et al. 2000). Anthonisen (1976) and Hellinga et al. (1998) reported HNO_2-N inhibited the nitrite oxidation at concentrations of 0.2–0.22 mg/L. Vadivelu et al. (2006) and Pratt et al. (2003) reported that free nitrous acid started inhibiting the anabolism of *Nitrobacter* at 0.011 mg HNO_2^-N/L (0.8 μM), and completely suppressed the biomass synthesis at 0.023 mg HNO_2^-N/L (1.6 μM).

Many lab-scale systems have achieved stable shortcut nitrification and denitrification through the inhibition of free ammonia (FA, NH_3-N). NOB are inhibited by NH_3-N in the range of 0.1–1.0 mg/L (Anthonisen 1976; Chang et al. 2002), while AOB can tolerate NH_3-N as high as 10–150 mg/L. However, FA only temporarily inhibits the activities of AOB and NOB (Anthonisen 1976; Peng et al. 2004b). The nitrite oxidation by NOB recovered when the FA concentration was lowered to 0.2 mg/L (Han et al. 2003). It should be noted that the FA concentration is affected by wastewater pH and temperature (Anthonisen 1976), which further affects the stability of shortcut nitrification (Fdz-Polanco et al. 1994; Cecen 1996; Cecen et al. 1996; Surmacz-Gorska et al. 1997).

Research Status and Unsolved Matters

A stable partial nitrification can be achieved by regulating one of the factors described above. DO concentration is an economically feasible control parameter. Low DO concentration will save aeration cost but may reduce COD biodegradation rate and cause sludge bulking. Furthermore, idiographic and practical conditions should be considered. For example, because of the high specific heat of water, it is impractical to raise wastewater temperature to facilitate AOB. It is necessary to consider the economic feasibility when using DO, temperature, pH, and inhibitor as control parameters.

SHARON is the first full-scale process in which nitrification/denitrification can be achieved with nitrite as the intermediate product (Hellinga et al. 1998). It has been used for treating sludge digestion liquid in Rotterdam, Dokhaven, Utrecht, Zwolle, and Beverwijk (all in The Netherlands). However, SHARON needs to be operated at high temperatures (>35°C) and high ammonium concentrations, which limit its application (STOWA 1995). In contrast, SBR systems with long sludge ages (>30 d) have successfully achieved partial nitrification at low temperatures (<13°C) when treating municipal wastewater (Peng and Zhu 2006).

Until now the most successful operation of partial nitrification via nitrite has been achieved in sequencing batch processes (Cecen 1996; Verstraete and Philips 1998; Hidaka et al. 2002). The only study for partial nitrification in a continuous-flow process was conducted by Schmidt et al. (2003) with influent NH_4^+-N higher than 50 mg/L. The current challenge is how to implement stable partial nitrification in continuous-flow processes treating wastewater with low ammonia concentration (<60 mg/L).

C. Anaerobic Ammonium Oxidation (ANAMMOX)

Mechanism and Advantages

The Anaerobic Ammonia Oxidation (ANAMMOX) process, developed at Delft University of Technology in the 1990s, is a novel and low-cost approach to removing nitrogen from wastewater (van Graaf et al. 1995; Strous et al. 1999; Fux et al. 2002). In ANAMMOX, ammonia is oxidized to nitrogen by anaerobic AOB with nitrite as the electron acceptor. Hydrazine and hydroxylamine are the intermediate products (Schalk et al. 1998; Jetten et al. 1999). External carbon sources are not needed in ANAMMOX because carbon dioxide serves as the main carbon source for anaerobic AOB (van Graaf et al. 1996). Equation 7 is the ANAMMOX reaction (Jetten et al. 1999):

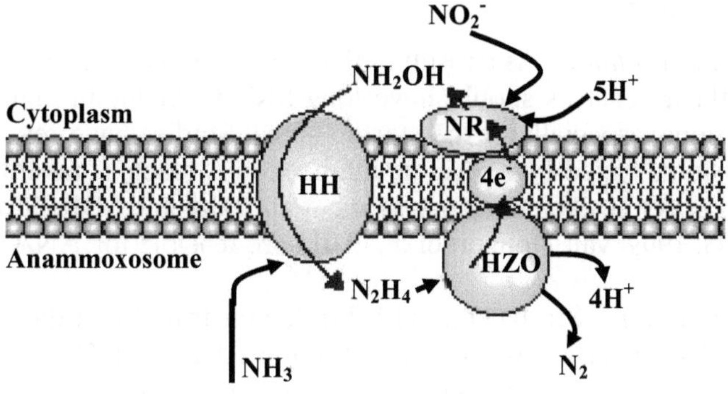

Fig. 5. Mechanism of anaerobic ammonium oxidation. NR is a nitrite-reducing enzyme (NH_2OH is the assumed product); HH (hydrazine hydrolase) condenses hydrazine from ammonia and hydroxylamine; HZO is a hydrazine-oxidizing enzyme (which may be equivalent to hydroxylamine oxidoreductase). (From Jetten et al. 2001.)

$$NH_4^+ + 1.31NO_2^- + 0.066HCO_3^- + 0.13H^+ \rightarrow$$
$$1.02N_2 + 0.26NO_3^- + 0.066CH_2O_{0.5}N_{0.15} + 2.03H_2O \quad (7)$$

The possible mechanism for anaerobic ammonium oxidation is shown in Fig. 5. Nitrite-reducing enzyme (NR) is on the cytoplasm side of the cell membrane. It catalyzes the reduction of NO_2^- to hydroxylamine. Hydrazine hydrolase (HH) across the cell membrane condenses hydroxylamine and ammonia to hydrazine. Hydrazine-oxidizing enzyme (HZO) is on the anammoxosome side of the cell membrane and catalyzes hydrazine to nitrogen. The electrons generated from these reactions are transferred back to NR.

Two ANAMMOX bacteria, tentatively named *Brocadia anammoxidans* (Strous et al. 1997a) and *Kuenenia stuttgartiensis* (Schmid et al. 2000; Cirpus et al. 2005), were found to carry out anaerobic ammonium oxidation. *Brocadia anammoxidans* was detected in the Netherlands, whereas *Kuenenia stuttgartiensis* was detected in Germany and Switzerland. These two bacteria have similar structures and produce hydrazine from exogenously supplied hydroxylamine. Two new species of ANAMMOX bacteria, *Candidatus Scalindua brodae* and *Candidatus Scalindua wagneri*, have been recently discovered (Schmid et al. 2003).

Compared with conventional nitrification-denitrification processes, ANAMMOX has two major advantages. First, because ANAMMOX is carried out by autotrophic bacteria, there is no need for organic carbon sources, which saves chemical dosage costs. Second, the biomass yield of ANAMMOX is very low (0.11 g VSS/g NH_4^+-N, VSS—volatile suspended solids), which saves sludge treatment costs (Jetten et al. 1999; Fux et al. 2002; Cirpus et al. 2005).

Key control factors

Reactor configuration. As a result of the slow growth rate of ANAMMOX bacteria, these reactors should have long SRT to maintain high biomass concentrations, especially at the startup stage. Studies have revealed that biofilm systems (fixed-bed reactor, fluidized-bed reactor, gas-lift reactor, etc.) (Sliekers et al. 2003; Dapena-Mora et al. 2004) and SBR (Strous et al. 1997c, 1998, 1999; van Dongen et al. 2001) are feasible for ANAMMOX.

DO concentration. Strous et al. (1997a) demonstrated that the activity of ANAMMOX bacteria was temporarily inhibited at the DO concentration of 0.2 mg/L and later recovered under anoxic conditions. The activity of ANAMMOX bacteria was completely inhibited at DO concentration of 0.2–1.0 mg/L.

Substrate concentration. Ammonia (the substrate for ANAMMOX) and nitrate (the by-product) produce little inhibition on ANAMMOX bacteria when their concentrations are below 1000 mg/L (Jetten et al. 1999). However, nitrite (another substrate) exhibits an adverse impact on ANAMMOX bacteria at a concentration of 100 mg/L (Strous et al. 1999). It is critical to maintain nitrite concentration below 70 mg/L in this process (Schmidt et al. 2002a).

pH. pH affects this process in terms of substrate constituents. The percentage of ammonia and nitrite in wastewater is significantly influenced by pH (Anthonisen 1976; Abeling and Seyfried 1992) and can be expressed in the following equations (Eqs. 8 and 9):

$$NH_3\,/\% = \frac{100}{1+\frac{10^{-pH}}{[K_a]}} \tag{8}$$

$$NHO_2\,/\% = \frac{100}{1+\frac{[K_a]}{10^{-pH}}} \tag{9}$$

in which $[K_a]$ is the ionization constant. Appropriate pH range for ANAMMOX bacteria is 7.7–8.3 with the maximum reaction rates occurring at pH of 8.0 (Strous et al. 1997a). Reaction rates increased at pH of 6.0–7.5 but decreased at pH 8.0–9.5.

Temperature. Temperature is an important factor for cell growth and metabolic activity. Normally, cells grow faster at higher temperature. Because the growth rate of ANAMMOX bacteria is very slow, there has been no accurate correlation between their growth rates and temperatures. The activation energy of ANAMMOX bacteria is similar to that of aerobic AOB (about 70 kJ/mol) (Strous et al. 1997b). ANAMMOX can take place at temperatures ranging from 6°C to 43 °C, whereas the optimal temperature for its bacteria is 26°–28°C (Fig. 6) (Thamdrup and Dalsgaard 2002). The reaction rate drops rapidly at temperatures lower than 15°C or higher than 40°C.

NO and NO_2. Both NO and NO_2 are the intermediate products of NOB. They affect not only ANAMMOX bacterial activities but also their growth rates. Schmidt et al. (2002b) found that consumption rates of NH_3 and NO_2^- and production rates of NO_3^- increased with addition of NO_2^-, and were highest at $[NO_2^-]$ of 50 mg/L, but dropped at $[NO_2^-]$ higher than 600 mg/L. The specific growth rate of *Brocadia anammoxidans* increased from 0.003 h^{-1} without the addition of NO_2^- to 0.004 h^{-1} at the $[NO_2^-]$ of 50 mg/L, but dropped to 0.0028 h^{-1} at $[NO_2^-]$ of 200 mg/L.

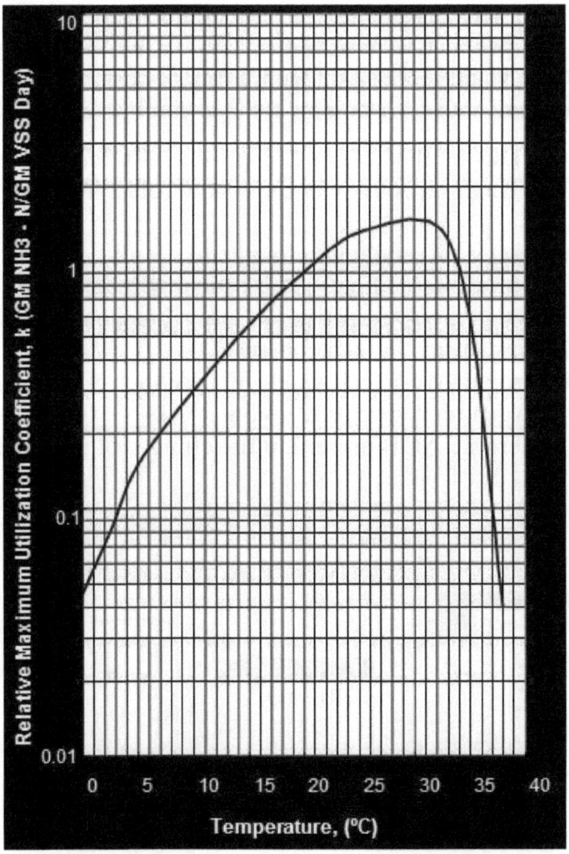

Fig. 6. Effect of temperature on growth rate of *ANAMMOX* bacteria.

Sludge age. Because of the slow growth rate and low biomass yield of ANAMMOX bacteria, a long sludge age is critical for this process. Although the theoretical doubling time of the bacteria is 11 d, longer SRT enhances ANAMMOX performance (Strous et al. 1999; Schmidt et al. 2002a).

Combined Partial Nitrification and ANAMMOX Process

The critical point for a successful process is to maintain a sufficient population of anaerobic AOB (Jetten et al. 1999). Because high [NO_2^-] inhibits anaerobic AOB, reducing NO_2^- accumulation will be a solution for this process. Partial nitrification can effectively convert NO_2^- to N_2 without significant accumulation; thereby, coupling with partial nitrification is expected to solve the NO_2^- problem in ANAMMOX (Fig. 7) (Fux et al. 2002). In this combined process, part of NH_4^+ is oxidized to NO_2^- by aerobic AOB, and NO_2^- is then reduced to N_2 by denitrifiers. The other part of NH_4^+ is oxidized to N_2 with NO_2^- as an electron acceptor by anaerobic AOB. The oxygen requirement of this combined process is 40% less than traditional nitrogen removal systems. The organic dosage for denitrification is also saved. In addition, sludge production is low because of the slow growth rate of anaerobic AOB, which reduces sludge treatment costs (Jetten et al. 1997). However, there are several problems for this combined system:

1. The residual DO in the effluent of partial nitrification might inhibit ANAMMOX bacterial activity because anaerobic AOB are sensitive to oxygen.
2. The optimal ratio of ammonia to nitrite should be 1.0:1.3 for the ANAMMOX process (see Eq. 5). This ratio might be difficult to maintain as a result of the involvement of complex biochemical reactions and diverse microorganisms in the process.
3. Because anaerobic AOB (cell yield, 0.11 g VSS/g NH_4^+-N) grow slower than aerobic AOB (cell yield, 0.13 g VSS/g NH_4^+-N), they will be outcompeted by the aerobic AOB present in the effluent of partial nitrification.

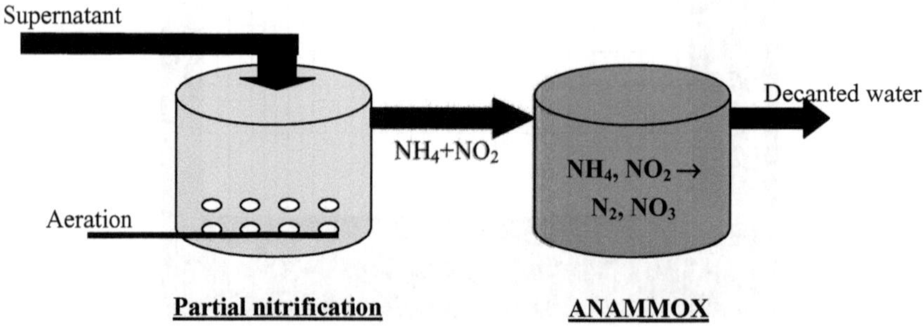

Fig. 7. The combined partial nitrification–*ANAMMOX* process.

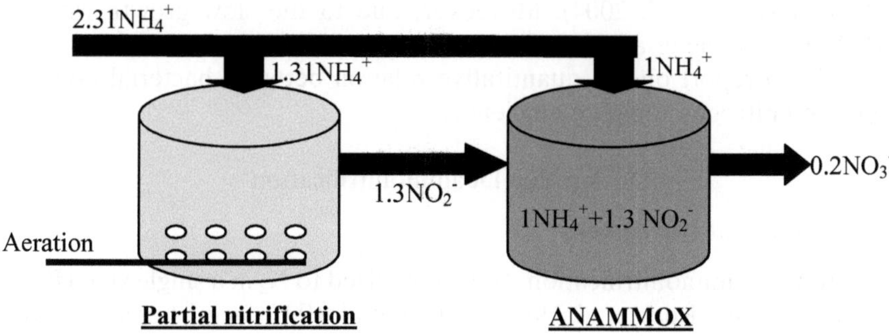

Fig. 8. Schematic graph of the combined partial nitrification–*ANAMMOX* process.

A step-feeding mode for partial nitrification and ANAMMOX can reduce the competition between anaerobic AOB and aerobic AOB (Fig. 8). Applying this operation mode can result in the following advantages:

1. Eliminating the competition of aerobic AOB and anaerobic AOB for ammonia, especially when ammonia concentration is low in raw wastewater.
2. Providing an optimal substrate ammonia and nitrite ratio (1.0:1.3) and thus enhancing total-N removal efficiency.
3. Ensuring an obligate anaerobic environment for ANAMMOX, because residual DO in the effluence of partial nitrification is depleted by COD in the influent.
4. Part of the nitrate produced in ANAMMOX can be removed by denitrification with the remaining COD as electron donors, which will improve total-N removal efficiency.
5. Because of the low growth rate of the bacteria, aerobic AOB in partial nitrification effluent might have a dilution or competition effect on ANAMMOX bacteria. The step-feeding strategy can reduce this dilution effect.

Research Status and Unsolved Matters

ANAMMOX has been operated in full-scale plants for treating sludge digestion supernatant in the Netherlands (Fux et al. 2002). Anammox effectively solves the nitrite inhibition problem in SHARON. The current challenge is how to efficiently accumulate anaerobic AOB. Strous et al. (1998) estimated that the cell yield value for anaerobic AOB was 0.066 mol cell/mol $(NH_4^+)_{reduced}$, ammonium consumption rate was 45 nmolNH_4^+/mgcell/min, and the maximum specific growth rate was 0.0027 hr^{-1}, which meant their doubling time was at least 11 d. Because the growth rate of these bacteria is slow, a long cell retention time is critical (Schmidt et al. 2003). When the level of cell loss is higher than cell growth, this process will become unstable

(Dapena-Mora et al. 2004). Moreover, due to the slow growth rate of ANAMMOX bacteria, it is difficult to analyze cell concentration. Until now there is no report on the quantitative relation between bacterial populations and nitrogen removal efficiency.

D. Aerobic Deammonitrification

Mechanism and Advantages

In aerobic deammonitrification, NH_4^+ is oxidized to N_2 in a single step (Poth and Focht 1985; Bock et al. 1995; Hippen et al. 1997; Siegrist et al. 1998b). Two models have been proposed for aerobic deammonitrification: the simultaneous nitrification and denitrification model (Fig. 9) and the separated nitrification and denitrification model (Fig. 10).

In the first model, aerobic deammonification is achieved by aerobic nitrifiers and anaerobic ANAMMOX AOB (Fig. 9). Alternate aerobic/anoxic biofilm reactors have been found to develop aerobic deammonification. Siegrist et al (1998a) observed that *Nitrosomomas* (aerobic AOB) on the surface layer of biofilm converted ammonia NH_4^+ to NO_2^- using oxygen from bulk wastewater in biological rotation contactors, and NH_4^+ and NO_2^- then transfused to the inner anoxic layer of biofilm and were removed by anaerobic AOB. This model has been verified stoichiometrically (Helmer-Madhok et al. 2002). *Paracoccus pantotropha* were found to carry out both anoxic and aerobic denitrification (Arts et al. 1995).

In the second model, aerobic deammonification is achieved by nitrifiers (mainly AOB, such as *Nitrosomomas*) (see Fig. 10). The AOB on the biofilm surface oxidize NH_4^+ to NO_2^- in the presence of oxygen. NO_2^- then transfuses

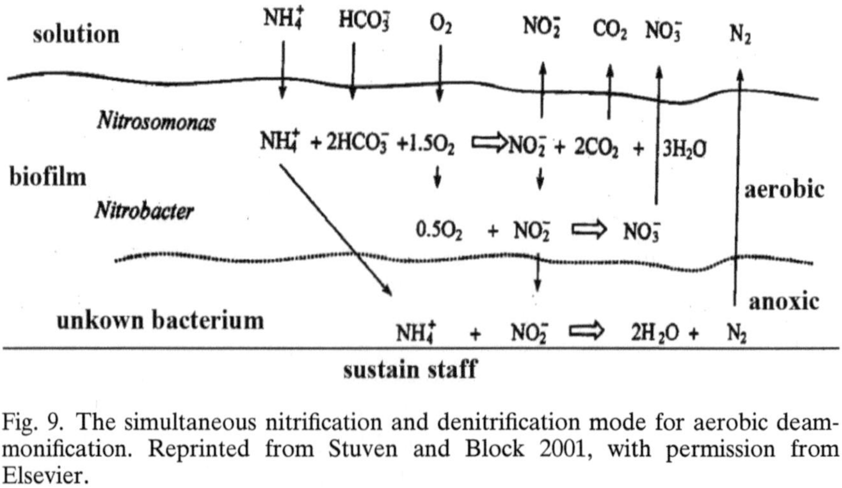

Fig. 9. The simultaneous nitrification and denitrification mode for aerobic deammonification. Reprinted from Stuven and Block 2001, with permission from Elsevier.

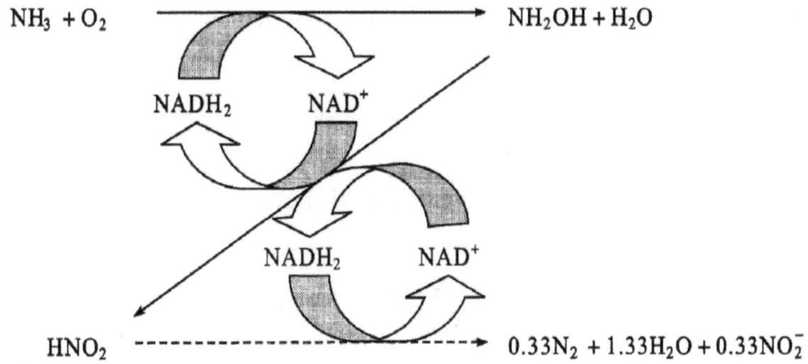

Fig. 10. Possible degradation of ammonia to dinitrogen and nitrite. Reprinted from Pochana and Keller 1999, with permission from IWA Publishing.

to the inner layer of the biofilm and is reduced to N_2 with $NADH_2$ as electron donor (Abeliovich 1987, 1992). In this model, hydroxylamine is the intermediate product of ammonium oxidation, and $NADH_2$ is the product of hydroxylamine oxidization. However, only 67% of NO_2^- can be converted to N_2 by AOB, based on this model (Siegrist et al. 1998b).

Research Status and Unsolved Matters

It has been found in pilot-scale and full-scale studies that NH_4^+ is oxidized to N_2 when treating municipal wastewater and landfill leachate (Siegrist et al. 1998a,b). Because aerobic deammonification normally occurs in the systems designed for conventional nitrification, the process design has not yet been optimized and nitrogen loading rates are low (90–250 g N/m^3 reactor d^{-1}) (Verstraete and Philips 1998). More studies need to be conducted to understand the mechanisms, characterize the microbial communities, and enhance process control.

E. Completely Autotrophic Nitrogen Removal over Nitrite (CANON)

Mechanism and Advantages

Because a significant amount of nitrogen is lost as nitrogen gas during the treatment of wastewater with high ammonia loadings but low organic loadings (Helmer and Kunst 1998; Koch et al. 2000; Helmer et al. 2001), a new process named Completely Autotrophic Nitrogen removal Over Nitrite (the CANON process) has been developed (Dijkman and Strous 1999). This process includes partial nitrification and anoxic oxidation of ammonia carried out by aerobic AOB and anaerobic AOB (Pynaert et al. 2002a,b; Third et al. 2001; Hao et al. 2002b; Nielsen et al. 2005). The interaction between these two types of nitrifiers under oxygen-limited conditions results in a complete conversion of ammonium to nitrogen gas in a single autotrophic reactor.

Under oxygen-limited condition, NH_4^+ is oxidized to NO_2^- by aerobic nitrifiers (Eq. 10) (Hanaki et al. 1990):

$$NH_4^+ + 1.5O_2 \rightarrow NO_2^- + 2H^+ + H_2O \quad (10)$$

Anaerobic AOB subsequently convert NH_4^+ and NO_2^- to nitrogen gas and NO_3^-:

$$NH_4^+ + 1.3NO_2 \rightarrow 1.02N_2^- + 0.26NO_3^+ + 2H_2O \quad (11)$$

With NO_2^- serving as the electron donor for the formation of biomass from carbon dioxide, the oxidation of NO_2^- to NO_3^- is stoichiometrically coupled with cell growth. The combination of reactions 9 and 10 results in the following overall nitrogen removal reaction (Eq. 12) (Strous 2000):

$$NH_4^+ + 0.85O_2 \rightarrow 0.435N_2 + 0.13NO_3^- + 1.3H_2O + 1.4H^+ \quad (12)$$

Both CANON and ANAMMOX removed nitrogen through the reaction of NH_4^+ and NO_2^-. However, NO_2^- is the electron donor in CANON and is produced in shortcut nitrification by AOB, whereas NO_2^- is the electron acceptor in ANAMMOX and needs to be added from other sources. CANON is operated at low oxygen condition (DO < 0.5 mg/L) whereas ANAMMOX is operated at obligate anaerobic condition.

CANON is cost-effective for wastewater with high ammonia concentrations (Pynaert et al. 2004). No extra carbon source is required, because it is completely autotrophic. In addition, nitrogen removal can be achieved in a single reactor with low aeration intensity. CANON consumes 63% less oxygen than conventional nitrogen removal processes (Sliekers et al. 2002).

Key Control Factors

There are three key factors for CANON: dissolved oxygen concentration, ammonia concentration, and an AOB population. Oxygen has two inhibition impacts on this process. It is toxic to anaerobic AOB and suppresses anaerobic AOB with the excessive production of nitrite. A DO concentration of 0.5 ± 0.07 mg/L is recommended (Sliekers et al. 2003).

Ammonia concentration is critical for this process. Ammonia oxidation is limited only by oxygen concentration when ammonia is sufficient in CANON, while nitrite oxidation is limited by both oxygen and nitrite concentrations. If either dissolved oxygen or nitrite is maintained at a low level, NOB can be inhibited in the system and ensure a stable CANON performance (Pynaert et al. 2002b; Nielsen et al. 2005). It has been found that a deficiency of ammonia substantially lowered this process efficiency, with 31% of NO_2^- generated by AOB reacting in anaerobic ammonification and 69% of nitrite reacting in nitrification. When ammonia became sufficient,

100% of NO_2^- generated by AOB reacted in anaerobic ammonification (Third et al. 2001). The experimental results showed that an ammonia loading of 14 mg/L hr provided a sufficient ammonia source in CANON (Third et al. 2001).

The interaction between aerobic AOB and anaerobic AOB affects this process (Third et al. 2001, 2005). These two types of bacteria live on different substrates, with aerobic AOB requiring ammonia and oxygen and anaerobic AOB requiring ammonia and nitrite. Therefore, CANON will be disrupted by the presence of NOB because NOB use oxygen and nitrite as substrates and compete with aerobic AOB for oxygen and with anaerobic AOB for nitrite.

Research Status and Unsolved Matters

The CANON process has been tested in full-scale nitrification systems (Helmer et al. 2001; Boran et al. 2004). N-removal rates reached 1.5 kg $N m^{-3} d^{-1}$ in this process (Sliekers et al. 2003), 20 times higher than other biological nitrogen removal processes (Kuai and Verstraete 1998; Sliekers et al. 2002). More studies are needed to clarify microbial communities in CANON, broaden its application, and enhance its resistance to shocks.

F. Oxygen-Limited Autotrophic Nitrification-Denitrification (OLAND)

Mechanism and Advantages

Nitrogen removal can also be accomplished in another single-step process, named the Oxygen-Limited Autotrophic Nitrification-Denitrification (OLAND) process (Kuai et al. 1998), in which AOB oxidize a portion of NH_4^+ to NO_2^- with oxygen as the electron acceptor and then reduce NO_2^- to N_2 with the other portion of NH_4^+ as the electron donor. OLAND is supposed to take place via two steps (reactions 13 and 14) (Poth 1986; Muller et al. 1995):

$$NH_4^+ + 1.5O_2 \rightarrow NO_2^- + H_2O + 2H^+ \quad (13)$$

$$NH_4^+ + NO_2^- \rightarrow N_2 + 2H_2O \quad (14)$$

Combining these two steps, we can get an overall reaction:

$$2NH_4^+ + 1.5O_2 \rightarrow N_2 + 3H_2O + 2H^+ \quad (15)$$

There is no clear distinction between OLAND and CANON. OLAND is achieved by aerobic AOB (*N. eutropha*) under oxygen-limited condition (Pynaert et al. 2004), while CANON is carried out by both aerobic AOB and anaerobic AOB under oxygen-limited condition. OLAND also exhibits a good tolerance of NH_4^+ and NO_2^- shocks (Windey et al. 2005). Compared with conventional nitrogen removal processes, OLAND consumes 63% less oxygen and does not require alkalinity dosage (Bock et al. 1995; Hippen et al. 1997; Pynaert et al. 2004).

Key Control Factors

Oxygen concentration is critical for OLAND because the population of aerobic AOB drastically decreases at low oxygen concentration (DO < 0.1 mgL^{-1}) (Kuai and Verstraete 1998; Philips et al. 2002; Zhang et al. 2004). Compared with CANON, OLAND has a shorter sludge retention time and a lower requirement for nitrite sources (Wyffels et al. 2004a; Windey et al. 2005).

Research Status and Unsolved Matters

Although OLAND is easier to operate than CANON, the application of this one-step process is severely limited by the low nitrogen removal efficiency (lower than 40%) and the uncertainty of the operational conditions (Schmidt and Bock 1997). To enhance OLAND performance, the oxidation rate of NH_4^+ to NO_2^- and the growth rate of aerobic AOB under oxygen-limited condition should be improved. In addition, the impacts of temperature and pH need to be studied, which is important for the growth of AOB and nitrogen removal efficiency.

G. Current Status of Nitrifying Bacteria

Understanding the population and function of nitrifying bacteria is critical to the design and operation of nitrogen removal processes. Because microorganisms are critical in nitrogen removal processes, many studies have investigated nitrifying bacterial species and activity. Based on gene sequence analysis, there are five genera of AOB: *Nitrosococcus, Nitrososmonas, Nitrosospira, Nitrosovibrio*, and *Nitrosolobus*; and four of NOB: *Nitrobacter, Nitrospira, Nitrococcus*, and *Nitrospira.* By using classical microbial screening techniques, and Painter (1986) found that *Nitrosomonas europa* and *Nitrobacter winogradskyi* were the main genera of AOB and NOB. In contrast, the studies using molecular biology-based techniques including 16S ribosomal RNA (rRNA)-targeted methods have shown a great diversity of nitrifiers in activated sludge. *Nitrosospira* and *Nitrospira* were found as the main genera of AOB and NOB in both bench-scale systems (Burrell et al. 1998; Schramm et al. 1998, 1999; Rittmann et al. 1999; Morgenroth et al. 2000; You et al. 2003; Gieseke et al. 2002) and wastewater treatment plants (Juretschko et al. 1998; Coskuner and Curtis 2002), whereas *Nitrosomonas* and *Nitrobacter* were still characterized as dominant nitrifying bacteria in some other studies using bench-scale systems (Gieseke et al. 2001; Chen et al. 2003: Tsuneda et al. 2003) and wastewater treatment plants (Wagner et al. 1996; Daims et al. 2001; Dionisi et al. 2002; Coskuner and

Curtis 2002; Hallin et al. 2005). In addition, although nitrifying bacterial populations (AOB + NOB) are generally supposed to be greater than 5%–8% in biomass for good nitrification (Randall et al. 1992; Koch et al. 2001), a wide variation in the percentage of nitrifying bacteria in the microbial community has been reported. It varied from 0.34% in activated sludge (Dionisi et al. 2002), through 6%–18% in a combined activated sludge and rotating biological contactor process (You et al. 2003) and a sewage plant (Wagner et al. 1995), and to more than 50% in a carbon-limited autotrophic nitrifying biofilm (Kindaichi et al. 2004) and an SBR system (Morgenroth et al. 2000). These studies reflect the differences between AOB and NOB populations in treatment facilities and raise two questions: (1) Does the dominance of specific nitrifying bacterial species vary with operational conditions and influent qualities? and (2) What is the correlation between microbial population and operational conditions in the treatment systems?

There are discrepancies in the dominance of AOB and NOB. It was generally accepted that AOB dominated over NOB with a ratio of 2.0–3.5 under conditions of good nitrification (Copp and Murphy 1995; You et al. 2003), possibly because of the inherent high growth rates of AOB (Schramm et al. 1999) and the higher energy generation of NH_4^+ oxidization by AOB than NO_2^- oxidization by NOB (You et al. 2003). Cell size could also contribute to the dominance of AOB, since AOB cells were larger than NOB (Altmann et al. 2003) and existed in larger colonies (Schramm et al. 1996; Juretschko et al. 1998). Another explanation was that AOB could maintain ribosome content under adverse conditions because inactive AOB could still be detected in high abundance by fluorescent *in situ* hybridization (FISH) targeting ribosomal RNA (rRNA) (Wagner et al. 1995; Okabe et al. 2004). However, several other studies revealed that the population of AOB was lower than that of NOB. Schramm et al. (1999) found that NOB (*Nitrospira*) was more than 30 times that of AOB (*Nitrosospira*) in a wastewater treatment plant, and Gieseke et al. (2001) found that NOB population (*Nitrospira*) was 85 times more than AOB (*Nitrosomonas*) in a bench-scale biofilm system. By using real-time quantitative PCR and assuming two copies of *amo*A gene per AOB cell and one copy of 16S rDNA gene per NOB cell, Dionisi et al. (2002) estimated the NOB population (*Nitrospira*) as 190 times greater than AOB (*Nitrosomonas*) in a wastewater treatment plant. This discrepancy between bacterial populations and their functions could be the result of different treatment processes and different types of genetic material targeted in these studies, and poses a requirement for information on gene expression or examination of the genetic material (e.g., mRNA) (Logan and Rittmann 1998).

To explain nitrifying bacterial communities in treatment systems, several studies assumed that *Nitrosomonas* and *Nitrobecter* were *r*-strategists (low affinity for substrates and high growth rates) and dominated at high substrate concentrations, whereas *Nitrosospira* and *Nitrospira* were *k*-strategists (high affinity for substrates and low growth rate) and thrived at low substrate concentrations (Manz et al. 1996; Schramm et al. 1998, 2000; Noguera et al. 2002). In addition, *Nitrobecter* was found to have the unique ability to live heterotrophically, whereas most of *Nitrospina, Nitrococcus*, and *Nitrospira* were unable to grow heterotrophically (Ehrich et al. 1995; Burrell et al. 1998).

Unsolved Matters

The importance of elucidating nitrifying bacterial populations and function in wastewater treatment processes has been well recognized. The combination of microelectrode and molecular biology has revealed the variation of different nitrifying bacterial species in the microenvironment of biofilm and activated sludge. However, there is one major unsolved problem: the application of microbial community in engineering design and operation. Although diverse nitrifying bacterial groups have been identified, the current activated sludge/biofilm models still assume nitrifying bacteria as a single group with the same cell growth rate and use empirical kinetic parameters, which leads to malfunction of the nitrogen removal process and the uncertainty of seeking the real causes. To enhance the performance of nitrogen removal processes, it is critical to incorporate the microbial community findings into the design and operation of treatment processes.

IV. Critical Comparisons of the Various Technologies

The novel and conventional nitrogen removal processes are compared in terms of reactor complexity, treatment performance, and operational costs (Table 2) (Jetten et al. 2002; Schmidt et al. 2003). From the aspect of reactor numbers, conventional technologies require two reactors while the novel technologies require only one, thus saving construction costs. As for oxygen requirement, conventional technologies normally require high dissolved oxygen concentrations to carry out complete nitrification, while the novel technologies need only low or limited oxygen supply, thus saving aeration costs. In terms of the addition of carbon sources, the conventional technologies require high influent C:N ratios for denitrification, while novel technologies have a low carbon requirement and exhibit good adaptation to both high and low C:N ratios.

With the low cell growth rates of anaerobic nitrifiers, novel technologies produce less sludge than conventional technologies and thus reduce sludge treatment costs. All novel technologies except ANAMMOX can efficiently remove nitrogen from municipal wastewater. These advantages make the novel technologies promising. However, several features of these technologies need to be improved, such as operational stability, nitrogen removal efficiency, and growth of specific nitrifiers.

Shortcut nitrification-denitrification (partial nitrification) and ANAMMOX are the most developed among these novel technologies.

Table 2. Comparison of Various Biological Nitrogen Removal Technologies and Processes.

Technology or process	Conventional	SND	Short-cut (SHARON)	ANAMMOX	Aerobic deammonitrification	CANON	OLAND
Number of reactors	2	1	2	1	1	1	1
Feed	Wastewater	Wastewater	Wastewater	Ammonia + Nitrite	Wastewater	Wastewater	Wastewater
Discharge	NO_3^-, NO_2^-, N_2	N_2	NO_2^-, N_2	NO_3^-, N_2	N_2	NO_3^-, N_2	N_2
Operating conditions	Aerobic, anoxic	Aerobic	Aerobic, anoxic	Anaerobic	Aerobic	Oxygen limited	Oxygen limited
Oxygen requirements	High	Low	Low	None	Low	Low	Low
Biomass retention	None	None	None	Yes	Yes	Yes	Yes
COD requirements	Yes	No	No	No	No	No	No
Sludge production	High	Low	Low	Low	Low	Low	Low
Bacteria	Nitrifiers + heterotrophs	Heterotrophic Nitrifiers + aerobic denitrifier	Aerobic Ammonium Oxidizer	Planctomycetes	Aerobic nitrifiers + aerobic Denitrifier	Aerobic ammonium Oxidizer + planctomycetes	Aerobic ammonium Oxidizer + anaerobic Ammonium oxidizer
Max N loading (kg N m^{-3} reactor d^{-1})	2–8	1–3.5	0.5–1.5	10–20	1–2	2–3	0.1
Total-N removal efficiency	95%	100%	90%	87%	60%	75%	85%
Optimum temperature (°C)	12–35	20–30	Above 25	30–40	Unknown	30–40	30–40
Common reactor configuration	Activated sludge and biofilm	Oxidation ditch, SBR	Activated sludge and biofilm	Fixed and fluidized-bed reactor, gas-lift reactor, SBR	Biological rotating contactor, gas-lift reactor, fixed and fluidized-bed reactor	Fixed and fluidized-bed reactor, SBR	Fixed and fluidized-bed reactor, SBR
Application status	Established	Laboratory	Full-scale plants	Full-scale initiated	Laboratory	Laboratory	Laboratory
Electron donor	COD	Unknown	COD	Ammonium	Ammonium	Nitrite	Ammonia
Biofilms or suspension	Biofilms/ suspension	Biofilms/ suspension	Suspension	Biofilms/ suspension	Biofilms/suspension	Biofilms/suspension	Biofilms/suspension

SND, simultaneous nitrification and denitrification; COD, Chemical oxygen demand; SBR, sequencing batch reactor.

Shortcut nitrification-denitrification achieves nitrogen removal in a single tank at a low DO concentration. ANAMMOX has high efficiency when operated at higher total-N loadings and can be combined with partial nitrification through anaerobic AOB. The SHARON and CANON processes are derived from these two technologies. Aerobic dammonitrification is a combined process of partial nitrification and ANAMMOX reactions occurring at different layers of biofilm.

Summary

This comprehensive review discusses diverse conventional and novel technologies for nitrogen removal from wastewater. Novel technologies have distinct advantages in terms of saving configuration, aeration, and carbon sources. Each novel technology possesses promising features and potential problems. For instance, SND and OLAND processes can achieve 100% total nitrogen removal, but the low oxygen concentration required by these two processes substantially reduces the nitrification rate, which limits their application. On the other hand, denitrification can still be carried out by aerobic denitrifiers at high DO levels in activated sludge process, but it is difficult to cultivate this type of bacteria.

The SHARON process is most commonly used for shortcut nitrification and denitrification because of its low requirements for retention time, oxygen concentration, and carbon source. However, its high operational temperature (about 35°C) limits the application. Several real-time control strategies (DO, pH, and ORP) have been developed to achieve a stable nitrite accumulation in SHARON.

The ANAMMOX process can sustain at high total-N loadings and has been employed in full-scale treatment plants, but the problem of nitrite supply has not been solved, and the treated wastewater still contains nitrate. In addition, the inoculation and enrichment of ANAMMOX bacteria (i.e., anaerobic AOB) is difficult. The problem of nitrite supply has been solved by combining partial nitrification with ANAMMOX, which provides abundant nitrite for anaerobic AOB. ANAMMOX is currently used for treating sludge digestion supernatant.

Aerobic dammonitrification is a process combining partial nitrification and ANAMMOX at different layers of biofilm. Although the technology has been tested in pilot- and full-scale experiments, the mechanism is still unclear.

CANON and OLAND are one-step ammonium removal processes that possess distinct advantages of saving carbon sources and aeration costs. The major challenge is the enrichment of anaerobic microorganisms capable of oxidizing ammonia with nitrite as the electron acceptor.

Molecular biology and environmental biotechnology can help identify functional microorganisms, characterize microbial communities, and develop new nitrogen removal processes. Extensive research should be conducted to apply and optimize these novel processes in wastewater treatment plants. More effort should be invested to combine these novel processes (e.g., partial nitrification, ANAMMOX) to enhance nitrogen removal efficiency.

Acknowledgments

This research is financially supported by National Natural Science Foundation Key International Cooperation project of China (No. 50521140075) and National Hi-Tech Development Program "863" project (No. 2004AA601020) and attached projects of "863" project of Beijing Municipal Science and Technology (Z0005186040421).

References

Abeling U, Seyfried CF (1992) Anaerobic–aerobic treatment of high strength ammonium wastewater nitrogen removal via nitrite. Water Sci Technol 26:1007–1015.

Abeliovich A (1987) Nitrifying bacteria in wastewater reservoirs. Appl Environ Microbiol 53:754–760.

Abeliovich A (1992) Transformations of ammonia and the environmental impact of nitrifying bacteria. Biodegradation 3:255–264.

Altmann D, Stief P, Amann R, de Beer D, Schramm A (2003) In situ distribution and activity of nitrifying bacteria in freshwater sediment. Environ Microbiol 5(9):798–803.

Anthonisen AC (1976) Inhibition of nitrification by ammonia and nitrous acid. J Water Pollut Control Fed 48(5):835–852.

Arts PAM, Robertson LA, Kuenen JG (1995) Nitrification and denitrification by *Thiosphaera pantotropha* in aerobic chemostat cultures. FEMS Microbiol Ecol 18(4):305–315.

Baeza JA, Gabriel D, Lafuente J (2004) Effect of internal recycle on the nitrogen removal efficiency of an anaerobic/anoxic/oxic (A^2/O) wastewater treatment plant (WWTP). Process Biochem 39:1615–1624.

Balmelle B (1992) Study of factors controlling nitrite build-up in biological processes for water nitrification. Water Sci Technol 26:1018–1025.

Baumann B, Snozzi M, Zehnder AJB, van der Meer JR (1996) Dynamics of denitrification activity of *Paracoccus denitrificans* in continuous culture during aerobic-anaerobic changes. J Bacteriol 178:4367–4374.

Beccari M, Marani E, Ramadori R, Tandoi V (1983) Kinetic of dissimilatory nitrate and nitrite reduction in suspended growth culture. J Water Pollut Control Fed 55:58–64.

Belser LW, Mays EL (1980) Specific inhibition of nitrite oxidation by chlorate and its use in assessing nitrification in soils and sediments. Appl Environ Microbiol 39:505–510.

Bock E, Schmidt I, Stüven R, Zart D (1995) Nitrogen loss caused by denitrifying *Nitrosomonas* cells using ammonium or hydrogen as electron donors and nitrite as electron acceptor. Arch Microbiol 163:16–20.

Boran K, van Niftrik L, Sliekers O, Schmid MC, Schmidt I, van de Pas-Schoonen K, Cirpus I, van der Star W, van Loosdrecht M, Abma W, Kuenen JG, Mulder J-W,

Jetten MSM, den Camp HO, Strous M, van de Vossenberg J (2004) Application, eco-physiology and biodiversity of anaerobic ammonium-oxidizing bacteria. Rev Environ Sci Bio/Technol 3:255–264.
Brouwer M, van Loosdrecht MCM, Heijnen JJ (1996) One reactor system for ammonium removal via nitrite. STOWA report 96-01. STOWA, Utrecht.
Bruce ER, Perry LM (2001) Environmental Biotechnology: Principles and Applications. McGraw-Hill, New York.
Burrell PC, Keller J, Blackall LL (1998) Microbiology of a nitrite-oxidizing bioreactor. Appl. Environ. Microbiol 64:1878–1883.
Camilla G, Guamelius L, Dalhammar G (1998) Comparison of inhibition assays using nitrogen removing bacteria: application to industrial wastewater. Water Res 32(10):2995–3000.
Castignetti D, Gunner HB (1981) Nitrite and nitrate synthesis from pyruvic oxime by an *Alcaligenes* sp. Curr Microbiol 5:379–384.
Castignetti D, Hollocher TC (1982) Nitrogen redox metabolism of a heterotrophic nitrifying-denitrifying *Alcaligenes* sp. from soil. Appl Environ Microbiol 44:923–928.
Cecen F, Orak E, Gokcin P (1995) Nitrification studies on fertilizer wastewaters in activated sludge and biofilm reactors. Water Sci Technol 32(12):141–148.
Cecen F (1996) Investigation of partial and full nitrification characteristics of fertilizer wastewaters in a submerged biofilm reactor. Water Sci Technol 34(11):77–85.
Chang JS, Cha GC, Kim DJ (2002) Nitrite accumulation characteristics in the nitrification of high strength ammonia wastewater with biofilm airlift suspension reactor. H K 40:114–120.
Chen GH, Wong MT, Okabe S, Watanabe Y (2003) Dynamic response of nitrifying activated sludge batch culture to increased chloride concentration. Water Res 37(13):3125–3135.
Chiou R, Ouyang C (2001) The effect of recycle ratio on nitrification removal in the combined predenitrification/nitrification biofilter system. J Chem Technol Biotechnol 76:559–564.
Chiu Y-C, Lee L-L, Chang C-N, Chao AC (2007) Control of carbon and ammonium ratio for simultaneous nitrification and denitrification in a sequencing batch bioreactor. Int Biodeterior Biodegrad. 59(1):1–7
Cirpus IEY, de Been M, Op den Camp JM, Strous M, Le Paslier D, Kuenen GJ, JeHen MSM (2005) A new soluble 10kDa monoheme cytochrome c-552 from the anammox bacterium Candidatus '*Kuenenia stuttgartiensis*'. FEMS Microbiol lett 252:273–278.
Copp JB, Murphy KL (1995) Estimation of the active nitrifying biomass in activated sludge. Water Res 29 (8):1855–1861.
Coskuner G, Curtis TP (2002). In situ characterization of nitrifiers in an activated sludge: detection of *Nitrobacter* spp. J Appl Microbiol 93: 431–437.
Daims H, Nielsen JL, Nielsen PH, Schleifer KH, Wagner M (2001). In situ characterization of *Nitrospira*-like oxidizing bacteria active in wastewater treatment plants. Appl Environ Microbiol 67:5273–5284.
Dapena-Mora A, Campos JL, Mosquera-Corral A, Jetten MSM, Méndez R (2004) Stability of the ANAMMOX process in a gas-lift reactor and a SBR. J Biotechnol 110:159–170.
de Beer D, Schramm A, Santegoeds MC, Nielsen KH (1998) Anaerobic processes in activated sludge. Water Sci Technol 37(4–5):605–608.

Dijkman H, Strous M (1999) Process for ammonia removal from wastewater. Patent 1999; PCT/NL99/00446.

Dionisi HM, Layton AC, Harms G, Gregory I, Robison K, Sayler G (2002) Quantificaiton of *Nitrosomonas ologitropha*-like ammonia-oxidizing bacteria and *Nitrospira* spp. from full-scale wastewater treatment plant. Appl Environ Microbiol 68:245–253.

Ehrich S, Behrens E, Lebedeva W, Ludwig W, Bock E (1995) A new obligatately chemolithotrophic, nitrite-oxidizing bacterium, *Nitrospira moscoviensis* sp nov. and its phylogenetic relationship. Arch Microbiol 164: 16–23.

EPA (1993) Nitrogen control. USEPA, Washington, DC.

Fdz-Polanco F, Villaverde S, Garcia PA (1994) Temperature effect on nitrifying bacteria activity in biofilters: activation and free ammonia inhibition. Water Sci Technol 30(11):121–130.

Fdz-Polanco F, Villaverde S, Garcia PA (1996) Nitrite accumulation in submerged biofilters: combined effects. Water Sci Technol 34(3–4):371–378.

Focht DD, Verstraete W (1977) Biological ecology of nitrification and denitrification. In: Alexander M (ed) Advances in Microbial Ecology. Plenum Press, New York, pp 135–214.

Fux C, Boehler M, Huber P, Brunner I, Siegrist H (2002) Biological treatment of ammonium-rich wastewater by partial nitrification and subsequent anaerobic ammonium oxidation (Anammox) in a pilot plant. J Biotechnol 99:295–306.

Fuerhacker M, Bauer H, Ellinger R, Sree U, Schmid H, Zibuschka F, Puxbaum H (2000) Approach for a novel control strategy for simultaneous nitrification and denitrification in activated sludge reactors. Water Res 34(9):2499–2506.

Gieseke A, Purkhold U, Wagner M, Amann R, Schramm A (2001) Community structure and activity dynamics of nitrifying bacteria in a phosphate-removing biofilm. Appl Environ Microbiol 67:1351–1362.

Gieseke A, Arns P, Amann R, Schramm A (2002) Simultaneous P and N removal in a sequencing batch biofilm reactor: insights from reactor-and microscale investigations. Water Res 36:501–509.

Gradly CPL, Lim H (1980) Biological Wastewater Treatment. Dekker, New York.

Guo HY, Zhou JT, Su J, Zhang ZY (2005) Integration of nitrification and denitrification in airlift bioreactor. Biochem Eng J 23:57–62.

Gupta AB (1997) *Thiosphaera pantotropha*: a sulphur bacterium capable of simultaneous heterotrophic nitrification and aerobic denitrification. Enzyme Microb Technol 21:589–595.

Hallin S, Lydmark H, Hermansson SF, Jarvis A, Kindgren P (2005) Community survey of ammonia-oxidizing bacteria in full-scale activated sludge processes with different solids retention time. J Appl Microbiol 99:629–640.

Han DW, Chang JS, Kim DJ (2003) Nitrifying microbial community analysis of nitrite accumulating biofilm reactor by Fluorescence in situ Hybridization. Water Sci Technol 47(1):97–104.

Hanaki K, Wantawin C, Ohgaki S (1990) Nitrification at low level of DO with and without organic loading in a suspended growth reactor. Water Res 24: 297–302.

Hao XD, Heijnena JJ, van Loosdrecht MCM (2002a) Model-based evaluation of temperature and inflow variations on a partial nitrification–ANAMMOX biofilm process. Water Res 36:4839–4849.

Hao XD, Heijnen JJ, van Loosdrecht MCM (2002b) Sensitivity analysis of a biofilm model describing a one–stage completely autotrophic nitrogen removal (CANON) process. Biotechnol. Bioeng 77:266–277.
Hellinga C, Schellen AAJC, Mulder JW, van Loosdrecht MCM, Heijnen JJ (1998) The SHARON process: an innovative method for nitrogen removal from ammonium-rich wastewater. Water Sci Technol 37(9):135–142.
Helmer C, Kunst S (1998) Simultaneous nitrification/denitrification in an aerobic biofilm system. Water Sci Technol 37:183–187.
Helmer C, Tromm C, Hippen A, Rosenwinkel KH, Seyfied CF, Kunst S (2001) Single stage biological nitrogen removal by nitrification and anaerobic oxidation biofilm systems. Water Sci Technol 43(1):311–320.
Helmer-Madhok C, Schmid M, Fillpov E, Gaul T, Hippen A, Rosewinkel H-H, Seyfried CF, Wagner M, Kunst S (2002) Deammonification in biofilm systems: population structure and function. Water Sci Technol 46(1–2):223–231.
Hibiya K, Terada A, Tsuneda S, Hirata A (2003) Simultaneous nitrification and denitrification by controlling vertical and horizontal microenvironment in a membrane-aerated biofilm reactor J Biotechnol 100:23–32.
Hidaka T, Yamada H, Kawamura M, Tsuno H (2002) Effect of dissolved oxygen conditions on nitrogen removal in continuously fed intermittent-aeration process with two tanks. Water Sci Technol 45:181–188.
Hippen A, Rosenwinkel KH, Baumgarten G, Seyfried CF (1997) Aerobic deammonification: a new experience in the treatment of wastewaters. Water Sci Technol 35(10):111–120.
Holman JB, Wareham DG (2005) COD, ammonia and dissolved oxygen time profiles in the simultaneous nitrification/denitrification process. Biochem Eng J 22:125–133.
Hu L, Wang J, Wen X, Qian Y (2005) Study on performance characteristics of SBR under limited dissolved oxygen. Process Biochem 40:293–296.
Hu TL, Kung KT (2000) Study of heterotrophic nitrifying bacteria from wastewater treatment systems treating acrylonitrile, butadiene and styrene resin wastewater, Water Sci Technol 42:3–4 315–322.
Hunik JH (1993) Engineering aspects of nitrification with immobilized cells. PhD thesis, Wageningen Agricultural University, The Netherlands.
Jetten MSM, Horn SJ, van Loosdrecht MCM (1997) Towards a more sustainable wastewater treatment system. Water Sci Technol 35:171–179.
Jetten MSM, Strous M, van de Pas-Schoonen KT, Schalk J, Udo van Dongen GJM, van de Graaf AA, Logemann S, Muyzer G, van Loosdrecht MCM, Kuenen JG (1999) The anaerobic oxidation of ammonium. FEMS Microbiol Rev 22:421–437.
Jetten MSM, Wagner M, Fuerst J, van Loosdrecht MCM, Kuenen G, Strous M (2001) Microbiology and application of the anaerobic ammonium oxidation ('anammox') process. Curr Opin in Biotechnol 12:283–288.
Jetten MSM, Schmid M, Schmidt I, Wubben M, van Dongen U, Abma W, et al (2002) Improved nitrogen removal by application of new nitrogen cycle bacteria. Rev Environ Sci BioTechnol 1:51–63.
Juretschko S, Timmermann G, Schmid M, Schleifer K, Roser PA, Koops H, Wagner M (1998) Combined molecular and conventional analyses of nitrifying bacterium diversity in activated sludge: *Nitrososcoccus mobilis* and *Nitrospira*-like bacteria as dominant populations. Appl Environ Microbiol 64(8):3042–3051.

Keller J, Subramaniam K, Gosswein J, Greenfield PF (1997) Nutrient removal from industrial wastewater using single tank sequencing batch reactors. Water Sci Technol 35:137–144.

Khin T, Annachhatre AP (2004) Novel microbial nitrogen removal processes. Biotechnol Adv 22:519–532.

Killham K (1986) Heterotrophic nitrification. In: Prosser JI (ed) Nitrification. IRL Press, Oxford, pp 117–126.

Kim DJ, Lee D-I, Keller J (2006) Effect of temperature and free ammonia on nitrification and nitrite accumulation in landfill leachate and analysis of its nitrifying bacterial community by FISH. Bioresour Technol 97:459–468.

Kim JK, Park KJ, Cho KS, Nam S-W, Park T-J, Rakesh B (2005) Aerobic nitrification-denitrification by heterotrophic *Bacillus* strains. Bioresour Technol 96:1897–1906.

Kindaichi T, Ito T, Okabe S (2004) Ecophysiological interaction between nitrifying bacteria and heterotrophic bacteira in autotrophic nitrifying biofilms as determined by microautoradiography-fluorescence in situ hybridization. Appl Environ Microbiol 70(3):1641–1650.

Koch G, Egli K, Van der Meer JR, Siegrist H (2000) Mathematical modeling of autotrophic denitrification in a nitrifying biofilm of a rotating biological contactor. Water Sci Technol 41:191–198.

Koch G, Kuhni M, Siegrist H (2001) Calibration and validation of an ASM3-based steady-state model for activated sludge systems. Part 1. Prediction of nitrogen removal and sludge production. Water Res 35:2235–2245.

Kshirsagar M, Gupta AB, Gupta SK (1995) Aerobic denitrification studies on activated sludge mixed with *Thiosphaera pantotropha*. Environ Technol Lett 16:35–43.

Kuai L, Verstraete W (1998) Ammonium removal by the oxygen-limited autotrophic nitrification-denitrification system. Appl Environ Microbiol 64(11):4500–4506.

Lee H-J, J Bae JH, Cho KM (2001) Simultaneous nitrification and denitrification in a mixed methanotrophic culture. Biotechnol Lett 23:935–941.

Li BK, Bishop P (2004) Micro-profiles of activated sludge floc determined using microelectrodes. Water Res 38:1248–1258.

Li DH, Ganczarczyk JJ (1990) Structure of activated sludge flocs. Biotechnol Bioeng 11:127–138.

Li DH, Ganczarczyk JJ (1993) Factors affecting dispersion of activated sludge flocs. Water Environ Res 65:258–263.

Logan BE, Rittmann BE (1998) Finding solutions for tough environmental problems. Environ Sci Technol 11:502–507.

Manz W, Amman R, Ludwig W, Vancanneyt M, Schleifer KHA (1996) Application of a suite of 16S rRNA-specific oligonucleotide probes designed to investigate bacteria of the phylum *Cytophage-Flavobacter-Bacteroides* in the natural environment. Microbiology 142:1097–1106.

Morgenroth E, Obermayer A, Arnold E, Bruhl A, Wagner M, Wilderer PA (2000) Effect of long-term idle periods on the performance of sequencing batch reactors. Water Sci Technol 41:105–113.

Mulder JW, van Loosdrecht MC, Hellinga C, van Kempen R (2001) Full-scale application of the SHARON process for treatment of rejection water of digested sludge dewatering. Water Sic Technol 43(11):127–134.

Muller EB, Stouthamer AH, van Verseveld HW (1995) Simultaneous NH_3 oxidation and N_2 production at reduced O_2 tensions by sewage sludge subcultured with chemolithotrophic medium. Biodegradation 6:339–349.
Munch EV, Lant PA, Keiler J (1996) Simultaneous nitrification and denitrification in bench-scale sequencing batch reactors. Water Res 30:277–284.
Neufeld R, Greenfield J, Rieder B (1986) Temperature, cyanide and phenolic nitrification inhibition. Water Res 20(5):633–642.
Nielsen M, Bollmann A, Sliekers O, Jetten M, Schmid M, Strous M, Schmidt I, Larsen LH, Nielsen LP, Revsbech NP (2005) Kinetics, diffusional limitation and microscale distribution of chemistry and organisms in a CANON reactor. FEMS Microbiol Ecol 51:247–256.
Noguera R, Melo L, Purkhold U, Wuertz S, Wagner M (2002) Nitrifying and heterotrophic population dynamics in biofilm reactors: effects of hydraulic retention time and the presence of organic carbon. Water Res 36:469–481.
Okabe S, Kindaichi T, Ito T, Satoh H (2004) Analysis of size distribution and cell density of ammonia-oxidizing bacterial microcolonies in relation to substrate microprofiles in biofilms. Biotechnol Bioeng 85 (1):86–95.
Østgaard K, Christensson M, Lie E, Jönsson K, Welander T (1997) Anoxic biological phosphorus removal in a full-scale UCT process. Water Res 31(11):2719–2726.
Pai TY, Tsai YP, Chou YJ, Chang HY, Leu HG, Ouyang CF (2004) Microbial kinetic analysis of three different types of EBNR process. Chemosphere 55: 109–118.
Painter HA (1986) Nitrification in the treatment of sewage and wastewater. In: Prosser JI (ed) Nitrification. IRL Press, Aberdeen, pp 185–211.
Patureau D, Godon JJ, Dabert P, Bouchez T, Bernet N, Delgenes JP, Moletta R (1998) *Microvirgula aerodenitrificans* gen. nov., sp. nov., a new gram-negative bacterium exhibiting co-respiration of oxygen and nitrogen oxides up to oxygen saturated conditions. Int J Syst Bacteriol 48:775–782.
Peng YZ, Zhu GB (2006) Biological nitrogen removal with nitrification and denitrification via nitrite pathway. Appl Microbiol Biotechnol 73(1):15–26.
Peng DC, Bernet N, Delgenes JP, Moletta R (2001) Simultaneous organic carbon and nitrogen removal in an SBR controlled at low dissolved oxygen concentration. J Chem Technol Biotechnol 76:553–558.
Peng YZ, Chen T, Tian WJ (2003) Nitrogen removal via nitrite at normal temperature in A/O process. J Environ Sci Health Part A Toxic-Hazard Subst Environ Eng 39(7):1667–1680.
Peng YZ, Chen Y, Peng CY, Liu M, Wang SY, Song XQ, Cui YW (2004a) Nitrite accumulation by aeration controlled in sequencing batch reactors treating domestic wastewater. Water Sci Technol 50(10):35–43.
Peng YZ, Li YZ, Peng CY, Wang SY (2004b) Nitrogen removal from pharmaceutical manufacturing wastewater with high concentration of ammonia and free ammonia via partial nitrification and denitrification. Water Sci Technol 50(6): 31–36.
Peng YZ, Yu DS, Liang DW, Zhu GB (2004c) Nitrogen removal via nitrite from seawater contained sewage. J Environ Sci Health Part A Toxic-Hazard Subst Envrion Eng 39(7):1667–1680.
Peng YZ, Zhu GB, Wang SY, Yu DS, Cui YW, Meng XS (2006) Pilot-scale studies on biological treatment of hypersaline wastewater at low temperature. Water Sci Technol 52:10–11.

Philips S, Wyffels S, Sprengers R, Verstraete W (2002) Oxygen-limited autotrophic nitrification/denitrification by ammonia oxidisers enables upward motion towards more favorable conditions. Appl Microbiol Biotechnol 59:557–566.

Picioreanu C, van Loosdrecht MCM, Heijnen JJ (1997) Modelling of the effect of oxygen concentration on nitrite accumulation in a biofilm airlift suspension reactor. Water Sci Technol 36:147–156.

Pochana K, Keller J (1999) Study of factors affecting simultaneous nitrification and denitrification (SND). Water Sci Technol 39(6):61–68.

Pochana K, Keller J, Paul L (1999) Model development for simultaneous nitrification and denitrification. Water Sci Technol 39(1):235–243.

Pollice A, Tandoi V, Lestingi C (2002) Influence of aeration and sludge retention time on ammonium oxidation to nitrite and nitrate. Water Res 36(10):2541–2546.

Poth M (1986) Dinitrogen production from nitrite by a *Nitrosomonas* isolate. Appl Environ Microbiol 51:957–959.

Poth M, Focht DD (1985) ^{15}N kinetic analysis of N_2O production by *Nitrosomonas europaea*: an examination of nitrifier denitrification. Appl Environ Microbiol 49:1134–1141.

Pratt S, Yuan Z, Gapes D, Dorigo M, Zeng RJ, Keller J (2003) Development of a novel titration and off-gas analysis (TOGA) sensor for study of biological processes in wastewater treatment systems. Biotechnol Bioeng 81:482–495.

Pynaert K, Sprengers R, Laenen J, Verstraete W (2002a) Oxygen-limited nitrification and denitrification in a lab-scale rotating biological contactor. Environ Technol 23(3):353–362.

Pynaert K, Sprengers R, Boeckx P, van Cleemput O, Verstraete W (2002b) Oxygen-limited nitrogen removal in a lab-scale rotating biological contactor treating an ammonium-rich wastewater. Water Sci Technol 45(10):357–363.

Pynaert K, Smets BF, Beheydt D, Verstraete W (2004) Startup of autotrophic nitrogen removal reactors via sequential biocatalyst addition. Environ Sci Technol 38:1228–1235.

Randall CW, Barnard JL, Stensel HD (1992) Design and Retrofit of Wastewater Treatment Plants for Biological Nutrient Removal. Technomic, Lancaster, PA.

Rick WY, Stuart MT (2001) Microbial nitrogen cycles: physiology, genomics and applications. Curr Opin Microbiol 4:307–312.

Rittmann BE, Laspidou CS, Flax J, Stahl DA, Urbain V, Harduin H, Waard J, Geurkink B, Henssen MJ, Brouwer H, Klapwijk A, Wetterauw M (1999). Molecular and modeling analyses of the structure and function of nitrifying activated sludge. Water Sci Technol 39 (1):51–59.

Robertson LA, Kuenen JG (1984) Aerobic denitrification: a controversy revived. Arch Microbiol 139:347–354.

Robertson LA, van Niel EWJ, Torresmans RAM, Kuenen JG (1988) Simultaneous nitrification and denitrification in aerobic chemostat cultures of *Thiosphaera pantotropha*. Appl Environ Microbiol 54:2812–2818.

Robertson LA, Dalsgaar T, Revsbech NP, Kuenen JG (1995) Confirmation of "aerobic denitrification" in batch cultures, using gas chromatography and 15N mass spectrometry. FEMS Microbiol Ecol 18:113–120.

Satoh H, Nakamura Y, Ono H, Okabe S (2003) Effect of oxygen concentration nitrification and denitrification in single activated sludge flocs. Biotechnol Bioeng 83:604–607.

Schalk J, Hege O, Kuene GJ, Jetten MSM (1998) The anaerobic oxidation of hydrazine: a novel reaction in microbial nitrogen metabolism. FEMS Microbiol Lett 158:61–67.

Schmid M, Twachtmann U, Klein M, Strous M, Juretschko S, Jetten MSM, Metzger JW, Schleifer KH, Wagner M (2000) Molecular evidence for genus level diversity of bacteria capable of catalyzing anaerobic ammonium oxidation. Syst Appl Microbiol 23:93–106.

Schmid M, Walsh K, Webb R, Irene W, Rijpstra C, van de Pas-Schoonen K, Verbruggen MJ, Hill T, Moffett B, Fuerst J, Schouten S, Damsté JSS, Harris J, Shaw P, Jetten M, Strous M (2003) *Candidatus* "*Scalindua brodae*," sp. nov., *Candidatus* "*Scalindua wagneri*," sp. nov. Two New Species of Anaerobic Ammonium Oxidizing Bacteria. Syst Appl Microbiol 26:529–538.

Schmidt I, Bock E (1997) Anaerobic ammonia oxidation with nitrogen dioxide by *Nitrosomonas eutropha*. Arch Microbiol 167:106–111.

Schmidt I, Sliekers O, Schmid M, Cirpus I, Strous M, Bock E, Kuenen JG, Jetten MSM (2002a) Aerobic and anaerobic ammonia oxidizing bacteria: competitors or natural partners. FEMS Microbiol Ecol 39:175–181.

Schmidt I, Hermelink C, van de Pas-Schoonen K, Strous M, Op de Camp HJ, Kuenen JG, Jetten MSM (2002b) Anaerobic ammonia oxidation in the presence of nitrogen oxides (NO_x) by two different lithotrophs. Appl Environ Microbiol 68(11):5351–5357.

Schmidt I, Sliekers O, Schmid M, Bock E, Fuerst J, Kuenen JG, Jetten MSM, Strous M (2003) New concepts of microbial treatment processes for the nitrogen removal in wastewater. FEMS Microbiol Rev 27:481–492.

Scholten E, Lukow T, Auling G, Kroppenstedt RM, Rainey FA, Dielmann H (1999) *Thaurea mecharnichensis* sp. nov., an aerobic denitrifier from a leachate treatment plant. Int J Syst Bacteriol 49:1045–1051.

Schramm A, Larsen LH, Revsbech PN, Ramsing BN, Amann R, Schleifer HK (1996) Structure and function of a nitrifying biofilm as determined by in situ hybridization and the use of microelectrodes. Appl Environ Microbiol 62:4641–4647.

Schramm A, de Beer D, Wagner M, Amann R (1998) Identification and activity in situ of *Nitrosospira* and *Nitrospira* spp. as dominant Populations in a nitrifying fluidized bed reactor. Appl Environ Microbiol 64:3480–3485.

Schramm A, de Beer D, Heuvel JC, Ottengraf S, Amann R (1999) Microscale distribution of populations and activities of *Nitrosospira* and *Nitrospira* spp. along a macroscale gradient in a nitrifying bioreactor: quantification by in situ hybridization and the use of microsensors. Appl Environ Microbiol 65(8):3690–3696.

Schramm A, de Beer D, Gieseke A, Amann R (2000) Microenvironments and distribution of nitrifying bacteria in a membrane-bound biofilm. Environ Microbiol 2:680–686.

Siegrist H, Reithaar S, Koch G, Lais P (1998a) Nitrogen loss in a nitrifying rotating contactor treating ammonium-rich wastewater without organic carbon. Water Sci Technol 38(8–9):241–248.

Siegrist H, Reithaar S, Lais P (1998b) Nitrogen loss in a nitrifying rotating contactor treating ammonium-rich leachate without organic carbon. Water Sci Technol 37(4–5):191–198.

Sliekers AO, Derwort N, Campos Gomez JL, Strous M, Kuenen JG, Jetten MSM (2002) Completely autotrophic ammonia removal over nitrite in one reactor. Water Res 36:2475–2482.

Sliekers AO, Third KA, Abma W, Kuenen JG, Jetten MSM (2003) CANON and Anammox in a gas-lift reactor. FEMS Microbiol Lett 218:339–344.
Snidaro D, Zartarian F, Jorand F, Bottero JY, Block JC, Manem J (1997) Characterization of activated sludge flocs structure. Water Sci Technol 36(4):313–320.
STOWA (1995) Treatment of nitrogen rich return flows of sewage treatment plants. Evaluation of Dutch pilot plant research projects (in Dutch). STOWA report, pp 95–08.
Strous M, van Gerven E, Kuenen JG, Jetten MSM (1997a) Effects of aerobic and microaerobic conditions on anaerobic ammonium-oxidating (Anammox) sludge. Appl Environ Microbiol 63(6):2446–2448.
Strous M, van Gerven E, Kuenen JG, Jetten M (1997b) Ammonium removal from concentrated waste streams with the Anaerobic Ammonium Oxidation (Anammox) process in different reactor configurations. Water Res 31:1955–1962.
Strous M, Van Gerven, Ping ZE, Kuenen JG, Jetten MSM (1997c) Ammonium removal from concentrated waste streams with the Anaerobic Ammonium Oxidation (Anammox) process in different reactor configurations. Water Res 31:1955–1962.
Strous M, Heijnen JJ, Kuenen JG, Jetten MSM (1998) The sequencing batch reactor as a powerful tool for the study of slowly growing anaerobic ammonium-oxidizing microorganisms. Appl Microbiol Biotechnol 50:589–596.
Strous M, Kuenen JG, Jetten MSM (1999) Key physiology of anaerobic ammonium oxidation. Appl Environ Microbiol 65:3248–3250.
Strous M (2000) Microbiology of anaerobic ammonium oxidation. PhD thesis, Kluyver Laboratory, TU Delft University, Delft, The Netherlands.
Stuven R, Bock E (2001) Nitrification and denitrification as a source for NO and NO_2 production in high-strength wastewater. Water Res 35(8):1905–1914.
Surmacz-Gorska J, Cichon A, Miksch K (1997) Nitrogen removal from wastewater with high ammonia nitrogen concentration via shorter nitrification and denitrification. Water Sci Technol 36(10):73–78.
Tam NFY, Wong YS, Leung G (1992) Effect of exogenous carbon sources on removal of inorganic nutrients by the nitrification denitrification process. Water Res 26:1229.
Tchobanoglous G, Burton FL, Stensel DH (2003) Wastewater Engineering: Treatment Disposal and Reuse, 4th Ed. McGraw-Hill, New York.
Thamdrup B, Dalsgaard T (2002) Production of N_2 through anaerobic ammonia oxidation coupled to nitrate reduction in marine sediments. Appl and Environ Microbiol 68(3):1312–1318.
Third KA, Sliekers AO, Kuenen JG, Jetten MM (2001) The CANON system (Completely Autotrophic Nitrogen-removal Over Nitrite) under ammonium limitation: interaction and competition between three groups of bacteria. Syst Appl Microbiol 24:588–596.
Third KA, Paxman J, Schmid M, Strous M, Jetten MS, Cord-Ruwisch R (2005) Enrichment of anammox from activated sludge and its application in the CANON process. Microb Ecol 49(2):236–244.
Tsuneda S, Nagano T, Hoshino T, Ejiri Y, Noda N, Hirata A (2003) Characterization of nitrifying granules produced in an aerobic upflow fluidized bed reactor. Water Res 37:4965–4973.
Turk O, Mavinic DS (1987) Benefits of using selective inhibition to remove nitrogen from highly nitrogenous wastes. Environ Technol Lett 8:419–426.

Turk O, Mavinic DS (1989) Maintaining nitrite build-up in a system acclimated to free ammonia. Water Res 23(11):1383–1388.

Vadivelu VM, Yuan Z, Fux C, Keller J (2006) The inhibitory effects of free nitrous acid on the energy generation and growth processes of an enriched *Nitrobacter* culture. Environ Sci Technol 40(14):4442–4448.

van Dongen U, Jetten MSM, van Loosdrecht MCM (2001) The SHARON-ANAMMOX process for treatment of ammonium rich wastewater. Water Sci Technol 44:53–60.

van Graaf AA, Mulder A, Bruijn P, Jetten MSM, Robertson LA, Kuenen JG (1995) Anaerobic ammonium oxidation is a biologically mediated process. Appl Environ Microbiol 61:246–1250.

van Graaf AA, Bruijn P, Robertson LA, Jetten MSM, Kuenen JG (1996) Autotrophic growth of anaerobic, ammonium-oxidising micro-organisms in a fluidized bed reactor. Microbiology (UK) 142:2187–2196.

van Niel EWJ (1991) Nitrification by heterotrophic denitrifiers and its relationship to autotrophic nitrification. PhD thesis, Delft University of Technology, Delft.

Verstraete W, Philips S (1998) Nitrification-denitrification processes and technologies in new contexts. Environ Pollut 102(S1):717–726.

Villaverde S, Garcia-Encina PA, Fdz-Polanco F (1997) Influence of pH over nitrifying biofilm activity in submerged biofilm. Water Res 31(5):1180–1186.

Villaverde S, Fdz-Polanco F, Garcia PA (2000) Nitrifying biofilm acclimation to free ammonia in the submerged biofilters. Start-up influence. Water Res 34:602–610.

Wagner M, Rath G, Amann R, Koops HP, Schleifer HK (1995) *In situ* identification of nitrifying ammonia-oxidizing bacteria. Syst Appl Microbiol 18:251–264.

Wagner MR, Koops G, Flood PH, Amann R (1996) In situ analysis of nitrifying bacteria in sewage treatment plants. Water Sci Technol 34 (1-2):237–244.

Wentzel MC, Ekama GA, Marais GVR (1992) Process and modelling of nitrification denitrification biological excess phosphorus removal systems: a review. Water Sci Technol 25(6):59–82.

Wilen BM, Balmer P (1999) The effect of dissolved oxygen concentration on the structure, size and size distribution of activated sludge flocs. Water Res 33(2): 391–400.

Williams SC, Beresford J (1998) The effect of anaerobic zone mixing on the performance of a three-stage Bardenpho plant. Water Sci Technol 38(1):55–62.

Windey K, De Bo I, Verstraete W (2005) Oxygen-limited autotrophic nitrification-denitrification (OLAND) in a rotating biological contactor treating high-salinity wastewater. Water Res 39(18):4512–4520.

Wouter G, Vandaele S, Verstraete W (1999) Nitrogen removal from sludge reject water with a membrane-assisted bioreactor. Water Res 33(1):23–32.

Wyffels S, Pynaert K, Boeckx P, Verstraete W, Van Cleemput O (2003) Identification and quantification of nitrogen removal in a rotating biological contactor by ^{15}N tracer techniques. Water Res 37:1252–1259.

Wyffels S, Boeckx P, Pynaert K, Zhang D, Van Cleemput O, Chen G, Verstraete W (2004a) Nitrogen removal from sludge reject water by a two-stage oxygen-limited autotrophic nitrification denitrification process. Water Sci Technol 49(5-6):57–64.

Wyffels S, van Hulle SWH, Boeckx P, Volcke EIP, van Cleemput O, van Rolleghem PA, Verstraete W (2004b) Modelling and simulation of oxygen-limited partial nitritafition in a membrane-assisted bioreactor (MBR). Biotechnol Bioeng 86:531–542.

Yoo H, Ann K-H, Lee H-J, Lee K-H, Kwak Y-J, Song K-G (1999) Nitrogen removal from synthetic wastewater by simultaneous nitrification and denitrification (SND) via nitrite in an intermittently-aerated reactor. Water Res 33(1):145–154.

You SJ, Hsu CL, Chuang SH, Ouyang CF (2003) Nitrification efficiency and nitrifying bacteria abundance in combined AS-RBC and A^2O systems. Water Res 37: 2281–2290.

Zhang D, Zhang DM, Liu YP, Cao WW, Chen GX (2004) Community analysis of ammonia oxidizer in the oxygen-limited nitritation stage of OLAND system by DGGE of PCR amplified 16S rDNA fragments and FISH. J Environ Sci China 16(5):838–842.

Zhang DJ, Lu PL, Long TR, Verstraete Willy (2005) The integration of methanogensis with simultaneous nitrification and denitrification in a membrane bioreactor. Process Biochem 40:541–547.

Zhao HW, Mavinic DS, Oldham WO, Koch FA (1999) Controlling factors for simultaneous nitrification and denitrification in a two-stage intermittent aeration process treating domestic sewage. Water Res 33:961–970.

Zheng P, Xu XL, Hu BL (2004) Novel Theories and Technologies of Biological Nitrogen Removal. Science Press, Beijing.

Zhu GB (2006) Performance and optimization of anoxic/oxic step feeding enhanced biological nitrogen removal process. PhD thesis, Harbin Institute of Technology, Harbin China.

Zhu GB, Peng YZ, Wu SY, Wang SY (2005) Automatic control strategy for step feed biological nitrogen removal process. J Environ Sci China 17(3):455–457.

Zhu GB, Peng YZ, Wu SY, Wang SY (2007a) Effect of influent flow rate distribution on the performance of step-feed biological nitrogen removal process. Chem Eng J 131:319–328.

Zhu GB, Peng YZ, Wang SY, Zuo JL, Wang YY, Guo JH (2007b) Development and experimental evaluation of a steady-state model for the step-feed biological nitrogen removal process. Chinese Journal of Chemical Engineering 15(3):411–417.

Manuscript received November 25, 2006; accepted February 1, 2007.

Index

MIX
Papier aus verantwortungsvollen Quellen
Paper from responsible sources
FSC® C105338

If you have any concerns about our products,
you can contact us on
ProductSafety@springernature.com

In case Publisher is established outside the EU,
the EU authorized representative is:
Springer Nature Customer Service Center GmbH
Europaplatz 3, 69115 Heidelberg, Germany

Printed by Libri Plureos GmbH
in Hamburg, Germany